ATOMIC AND ION COLLISIONS IN SOLIDS AND AT SURFACES

This book is an introduction to the application of computer simulation and theory in the study of the interaction of energetic particles (<1 eV to the mega-electronvolt range) with solid surfaces.

The authors describe methods that are applicable both to hard collisions between nuclear cores of atoms and to soft interactions, in which chemical effects or long-range forces dominate. The range of potential applications of the technique is enormous. In surface science, applications include surface atomic structure determination using ion scattering spectroscopy or element analysis using SIMS or other techniques that involve depth profiling. Industrial applications include optical or hard coating deposition, ion implantation in semiconductor device manufacture and nanotechnology. Plasma–sidewall interactions in fusion devices may also be studied using the techniques described.

This book will be of interest to graduate students and researchers, both academic and industrial, in surface science, semiconductor engineering, thin-film deposition and particle–surface interactions in departments of physics, chemistry and electrical engineering.

ATOMIC AND ION COLLISIONS IN SOLIDS AND AT SURFACES

THEORY, SIMULATION AND APPLICATIONS

Edited by Roger Smith

ROGER SMITH

Loughborough University

MARIO JAKAS

Universidad de la Laguna

DAVE ASHWORTH, BOB OVEN AND MARK BOWYER

University of Kent

IVAN CHAKAROV AND ROGER WEBB

University of Surrey

CAMBRIDGE
UNIVERSITY PRESS

PHYS
.08695726

PUBLISHED BY THE PRESS SYNDICATE OF THE UNIVERSITY OF CAMBRIDGE
The Pitt Building, Trumpington Street, Cambridge CB2 1RP, United Kingdom

CAMBRIDGE UNIVERSITY PRESS
The Edinburgh Building, Cambridge CB2 2RU, United Kingdom
40 West 20th Street, New York, NY 10011-4211, USA
10 Stamford Road, Oakleigh, Melbourne 3166, Australia

© Cambridge University Press 1997

First published 1997

Printed in the United Kingdom at the University Press, Cambridge

Typeset in Times 11/14pt

A catalogue record of this book is available from the British Library

Library of Congress cataloguing in publication data applied for

ISBN 0 521 44022 x hardback

Contents

Contents

1

Introduction

The purpose of this book is to give an introduction to some of the non-experimental techniques available for studying the interaction of energetic particles with solid surfaces. By energetic we mean particles with energies from <1 eV up to the mega-electronvolt range. The word non-experimental is chosen carefully because much of the book focuses on computer simulation in addition to basic theory. Simulation is a relative scientific newcomer, which contains elements both of theory and of experiment within its borders. A simulation is not a theory but a numerical model of a system. If it is a good model one may explore the behaviour of the real system by changing the numerical value of its input parameters and noting the changed responses. Simulations enable one to determine which are the important factors in a physical system that control its behaviour without the need necessarily to perform complex and expensive experiments. Sometimes we can probe areas that no experiment can determine, for example, the displacement and mixing of identical atoms in an atomic collision cascade. Usually, in performing the computational experiments on a model, the important parameters should be identified and need to be fixed at the start of the calculations. Usually we perform a sensitivity analysis by varying one parameter at a time.

This book is intended to describe methods that will be applicable both to hard collisions between nuclear cores of atoms and to soft interactions in which chemical effects or long-range forces dominate. Such an ability to model the effect of chemical reactions and the similarity between chemical processes and the motion of the planets were predicted many years ago by the eighteenth century philosophers. A. L. Lavoisier postulated in 1782 that 'Perhaps the precision of data will be brought to the point where the geometer will be able to calculate the phenomenon of any chemical combination whatsoever in the same way as (s)he calculates the motion of celestial bodies' and P. J. Macquer stated in 1771 that 'A man sufficiently

1

intelligent in both (chemistry and mathematics) might lay the foundation for a new science or rather might render the application of algebra and geometry to natural philosophy more general'. So the basic philosophy of using theory and simulation to study atomic interaction phenomena was foreseen over two centuries ago even if the techniques for performing the calculations were then unavailable.

For the most part, this book uses a classical dynamics description of the atomic and molecular interaction phenomena. The techniques described in this book are aimed at graduate students and surface scientists without necessarily a strong theoretical background. Thus we have avoided a strictly theoretical approach as might be found in the book by Mott and Massey (1965) on atomic collision phenomena. Where theory is included we have tried to derive the results from first principles wherever possible or referenced original articles. Thus Chapter 2 develops the basic equations used for modelling two-body collisions, whereas Chapter 5 develops the transport theory approach of linear collision cascades. Chapter 7 discusses the binary collision algorithm and Chapter 8 molecular dynamics computer models.

We define a *binary collision* to be a collision between two particles alone. Of course in a solid this is an approximation because no collision is ever really complete before the next collision begins, so we can modify the definition to be *a collision which is well separated from any other possible interaction involving the two collisional partners during the time of their interaction*. If a cascade is modelled by assuming that the collisions are binary in nature, then it is possible to develop both an analytic theory of the cascade and some fast simulation codes for their analysis. Otherwise the simulation has to solve the full equations of motion. Computer simulations that solve Newton's equations on a time step basis are usually called *molecular dynamics* or sometimes *multiple-interaction* simulations. They were first carried out in the 1950s by Alder and Wainright (1957) to study phase transitions. Such simulations applied to atomic collisions have shown that there are many simultaneous interactions between moving particles and so the assumption of a cascade comprised of binary collisions is at best a useful approximation. The BC approximation does have the advantage of allowing the determination of the particle trajectory asymptotes before and after an encounter without the determination of the intermediate dynamics. A simulation code based on such an approach goes from event to event skipping the intermediate dyamics and so is much faster to implement than a full MD simulation. Such programs are sometimes referred to as *event store* programs as opposed to a *time step* program, which is the basis of the MD simulations.

The basic assumption throughout the book is that *the system may be treated entirely by classical dynamics* once the interatomic potentials and inelastic loss mechanisms have been defined. (The interaction energies of the particles and ion beams are such that the velocities involved are non-relativistic.) In addition, the de Broglie wavelength is small. For example, the wavelength for an Ar atom at 100 eV is of the order of 0.005 Å. At room temperature the thermal vibration of the atoms in a crystal lattice is several hundred times greater than this, for example the typical vibration amplitude for Cu at room temperature is 0.1 Å. Quantum mechanics enters the discussion only in the form of the interatomic potential function. These potential functions are assumed to be independent of the relative kinetic energy of the interacting particles and usually are assumed to be unchanged by excitation or ionisation processes that may occur in a collision, although it would be relatively straightforward to develop such a model in a computer simulation. Indeed, recent work has indicated that dynamical computer simulations based on quantum theory are now feasible. Fully quantum mechanical *ab initio* molecular dynamics using density functional theory to calculate the interatomic forces have been carried out (Car and Parrinello, 1985). These are generally only applicable for sub-electronvolt energies and for ground state atoms, so there is some way to go before they can be used to model an entire atomic collision cascade. Currently (1995) it is possible with the aid of powerful parallel computers to deal with such systems of a few hundred atoms. Some interesting results of interest in catalysis are beginning to appear in the literature, such as the break-up of Cl_2 particles hitting a Si surface (Gillan, 1993). Similar studies can also be carried out using empirical potentials to fit material properties. For best results these are usually many-body in nature and can predict well many observed phenomena. It ought to be feasible soon to test such potentials against results obtained from a smaller system using the full quantum mechanical treatment. By comparison the largest MD simulations with simple potentials on parallel machines in 1993 have involved as many as 250 000 000 particles (Lomdahl *et al.*, 1993).

Visualisation is becoming an increasingly important aspect of MD computer simulation. Although cost prevents the publication of colour plates in the book, the use of three-dimensional colour graphical representations of atomic and molecular structures will be of increasing importance as a diagnostic tool. Moving picture representations of dynamic simulations can now be produced with relative ease and can be run on most desktop computers or captured on video. Some journals now accept computer-generated videos in addition to written articles. Figure 1.1 is used to illustrate the power of graphical visualisation achieved using the public domain package 'Rayshade' available free of charge for most Unix workstations (Kolb, 1992)

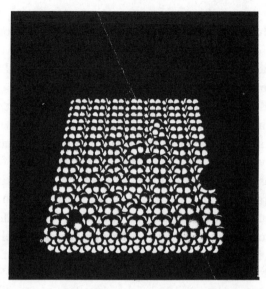

Fig. 1.1 Surface damage on Si{100}(2 × 1) induced by normal incidence 3 keV Ar$^+$ ions.

and shows a computer simulation of surface damage induced on the dimer reconstructed Si{100}(2 × 1) face as a result of normal incidence bombardment by 3 keV Ar$^+$ ions. Similar damage has been observed in the scanning tunnelling microscope.

The pioneering work of Jens Lindhardt and co-workers following in the Danish tradition of Niels Bohr has formed the basis of the development of transport theory for the analysis of atomic collision cascades over the years. Particular applications of the theory from the initial work on stopping powers and ranges through to specific applications such as the analysis of sputtering and mixing have been developed by other authors such as Sanders, Winterbon, Sigmund and Gras-Marti. Transport theory is included in this book in a self-contained chapter and formed the basis of many of the early developments in atomic collision theory. However, the statistical averaging required means that it is generally only useful for linear cascades (those in which different branches of a developing cascade do not overlap) in amorphous or polycrystalline materials and computer simulation is finding increasing favour amongst experimentalists because of its ease of use and generally more widespread applicability. For example, all the results of transport theory concerning ranges of energetic particles in matter can be reproduced using a binary collision computer code such as TRIM (Biersack and Eckstein, 1984) or MARLOWE (Robinson, 1989). However, the simulations can generally provide more information enabling the analysis of more complex systems even on a small desktop computer, for example atomic collisions in layered materials. Transport models on the other hand can have the advantage of providing very simple formulae for observed quantities and often can give useful physical insight into the problem.

The BC codes have a proven track record over many years for the simulation of implantation depths of particles in the kilo- to mega-electronvolt range. Originally these programs were effectively the same as a Monte Carlo solution to the transport equation but the target crystal structure was introduced by Robinson and with this the discovery of channelling. Today the MARLOWE codes of Robinson for crystalline materials and the TRIM codes of Biersack for amorphous materials are widely used in most surface science laboratories.

The MD simulations of radiation damage were pioneered by Vineyard's group at Brookhaven in the 1960s (Gibson *et al.*, 1960) and later developed extensively for sputtering studies by Don Harrison (Harrison, 1988). It was the application of MD to sputtering that first began to interest the surface science community with the simulation of the ejection patterns, the so-called *Wehner spots* first observed experimentally in the 1950s and the analysis of cluster ejection which is of particular interest to the SIMS community.

By way of conclusion to this introduction, the authors would like to give some justification for the book. It was felt that it was important that a book be written that contained an introduction to all the available techniques of analysis of atomic collision phenomena. Thus, although the lecture notes by Sigmund (1972) contain a useful introduction to transport theory, the book by Eckstein (1991) an important contribution on the BC codes and a very useful bibliography of many simulation papers and the book by Allen and Tildersley (1987) a good description of many aspects of MD, no book has attempted to develop all three approaches and to contrast them. In addition, the simulation field is developing very quickly with increasing computing power. MD simulations are now as fast as Monte Carlo simulations were ten years ago, so that many experimental scientists are beginning to use simulation tools as a matter of course. No doubt many new results will appear before this book has gone to press.

References

Allen, M. P. and Tildersley, D. J. (1987). *Computer Simulation of Liquids*, Clarendon Press, Oxford.
Alder, B. J and Wainwright, T. E. (1957). *J. Chem. Phys.* **27** 1208.
Biersack, J. P. and Eckstein, W. D. (1984). *Appl. Phys.* A **34** 73.
Car, R. and Parrinello, M. (1985). *Phys. Rev. Lett.* **35** 2471.

Eckstein, W. D. (1991). *Computer Simulation of Ion-Solid Interactions*, Springer-Verlag, Berlin.

Gibson, J. B., Goland, A. N., Milgrim, M. and Vineyard, G. H. (1960). *Phys. Rev.* **120** 1299.

Gillan, M. (1993). *New Scientist*, 3 April, p35.

Harrison, D. E. (1988). *CRC Crit. Rev. Solid St. Mater. Sci.* **14** S1.

Kolb, C. (1992). *Rayshade Manual*, Computer Science Department, Princeton University, Princeton, NJ 08544, USA.

Lavoisier, A. L. (1782). Mémoire sur l'affinité du Principie oxygène, *Mém. Acad. R. Soc.* 530.

Lomdahl, P., Beazley, D. M., Tamayo, P. and Gronbech-Jensen, N. (1993). *Int. J. Mod. Phys.* C **4** 1075.

Macquer, P. J. (1771). *A Dictionary of Chemistry* Vol. 1, London, 324.

Mott, N. F. and Massey, H. F. W. (1965). *The Theory of Atomic Collisions*, (3rd edition), Oxford University Press, Oxford.

Robinson, M. T. (1989). *Phys. Rev.* B **40** 10717.

Sigmund, P. (1972). *Rev. Roum. Phys.* **17** 823, 969, 1079.

2

The binary collision

2.1 Fundamentals

In this section we develop the theory for the two-body scattering problem. Many computer simulation programs treat the atomic collision problem by assuming that the particles interact in a pairwise fashion. Such a pairwise interaction is known as a binary collision. Binary collision programs are written with the assumption that the particle paths can be replaced by their asymptotes before and after collision. Thus it is necessary to be able to determine the equations for these asymptotes and the velocities of the particles after the collision. To determine these quantities will be one of the main tasks of this chapter. The starting point is the problem of scattering in a central force field.

2.1.1 Scattering by a central force field

We first derive expressions for the velocity and acceleration in spherical co-ordinates. Spherical co-ordinates (r, θ, ϕ) are related to Cartesian co-ordinates (x, y, z) by

$$r = x\mathbf{i} + y\mathbf{j} + z\mathbf{k} = r \sin \theta \cos \phi \mathbf{i} + r \sin \theta \sin \phi \mathbf{j} + r \cos \theta \mathbf{k}.$$

Here $\mathbf{i}, \mathbf{j}$ and $\mathbf{k}$ are unit vectors parallel to the x, y and z axes, $r = |\mathbf{r}|, \theta$ is the polar angle and ϕ the azimuthal angle.

It is also possible to define unit vectors $(\hat{r}, \hat{\theta}, \hat{\phi})$ parallel to the directions of increasing (r, θ, ϕ). These unit vectors are defined by

$$\hat{r} = \frac{\partial \mathbf{r}}{\partial r} \bigg/ \left| \frac{\partial \mathbf{r}}{\partial r} \right|, \quad \hat{\theta} = \frac{\partial \mathbf{r}}{\partial \theta} \bigg/ \left| \frac{\partial \mathbf{r}}{\partial \theta} \right|, \quad \hat{\phi} = \frac{\partial \mathbf{r}}{\partial \phi} \bigg/ \left| \frac{\partial \mathbf{r}}{\partial \phi} \right|.$$

Thus

$$\hat{r} = \sin\theta\cos\phi\, i + \sin\theta\sin\phi\, j + \cos\theta k,$$
$$\hat{\theta} = \cos\theta\cos\phi\, i + \cos\theta\sin\phi\, j - \sin\theta k, \qquad (2.1)$$
$$\hat{\phi} = -\sin\phi\, i + \cos\phi\, j.$$

Differentiating (2.1) gives

$$\frac{d\hat{r}}{dt} = \dot{\theta}\hat{\theta} + \sin\theta\dot{\phi}\hat{\phi},$$
$$\frac{d\hat{\theta}}{dt} = -\dot{\theta}\hat{r} + \cos\theta\dot{\phi}\hat{\phi}, \qquad (2.2)$$
$$\frac{d\hat{\phi}}{dt} = -\dot{\phi}\sin\theta\hat{r} - \dot{\phi}\cos\theta\hat{\theta},$$

where the overdot means differentiated with respect to time.

In spherical co-ordinates, the position vector r can be written

$$r = r\hat{r},$$

so differentiating and using (2.2)

$$\dot{r} = \dot{r}\hat{r} + r\dot{\theta}\hat{\theta} + r\sin\theta\dot{\phi}\hat{\phi},$$
$$\ddot{r} = \left(\ddot{r} - r\dot{\theta}^2 - r\sin^2\theta\dot{\phi}^2\right)\hat{r} + \left(r\ddot{\theta} + 2\dot{r}\dot{\theta} - r\sin\theta\cos\theta\dot{\phi}^2\right)\hat{\theta} \qquad (2.3)$$
$$+ \left(2\sin\theta\dot{r}\dot{\phi} + 2r\cos\theta\dot{\theta}\dot{\phi} + r\ddot{\phi}\sin\theta\right)\hat{\phi}.$$

A central force field at the origin 0 is defined as one such that the force is always along the radial vector from 0.

$$F = F\hat{r},$$

so from Newton's second law of motion

$$m\ddot{r} = F\hat{r}$$

and taking the vector product of both sides with r gives

$$r \times m\ddot{r} = F\left(r \times \hat{r}\right) = 0.$$

Similarly, since $\dot{r} \times \dot{r} = 0$

$$\frac{d}{dt}(r \times m\dot{r}) = r \times m\ddot{r} = 0. \qquad (2.4)$$

The quantity

$$L = r \times m\dot{r} \qquad (2.5)$$

is the angular momentum of the particle and the integral of (2.4) shows that

L is a constant vector in a central force field. Equation (2.5) shows that one possibility, $L = 0$ requires that $\dot{r}$ be parallel to r i.e. rectilinear motion $\left(\dot{\theta} = \dot{\phi} = 0\right)$. We can determine L in spherical or Cartesian co-ordinates using (2.3) and (2.2)

$$L = -mr^2 \sin \theta \dot{\phi} \hat{\theta} + mr^2 \dot{\theta} \hat{\phi}$$
$$= -mr^2 \left(\sin \theta \cos \theta \cos \phi \dot{\phi} + \sin \phi \dot{\theta}\right) i + mr^2 \left(\cos \phi \dot{\theta} \right. \tag{2.6}$$
$$\left. - \sin \theta \cos \theta \sin \phi \dot{\phi}\right) j + mr^2 \sin^2 \theta \dot{\phi} k.$$

Since L is a constant vector we can choose the axes so that L is parallel to k. This condition is achieved when $\dot{\theta} = 0$ and $\theta = \pi/2$. The central force problem then becomes motion in the x–y plane.

With $\theta = \pi/2, \dot{\theta} = 0$ and $L = mr^2 \dot{\phi} k$ equations (2.3) become

$$\dot{r} = \dot{r}\hat{r} + r\dot{\phi}\hat{\phi}, \tag{2.7a}$$
$$\ddot{r} = \left(\ddot{r} - r\dot{\phi}^2\right)\hat{r} + \left(2\dot{r}\dot{\phi} + r\ddot{\phi}\right)\hat{\phi}, \tag{2.7b}$$
$$\dot{\phi} = \frac{L}{mr^2}. \tag{2.7c}$$

If the central force $F(r)$ is written as the gradient of a potential function V then (2.7b) gives

$$m\left(\ddot{r} - r\dot{\phi}^2\right) = -\frac{\partial V}{\partial r}, \tag{2.7d}$$

which integrates to give the energy conservation law

$$\frac{1}{2}m\left(\dot{r}^2 + r^2\dot{\phi}^2\right) + V(r) = \text{ constant } = E_0 \tag{2.8}$$

thus

$$\dot{r} = \pm\left\{2/m\left[E_0 - V(r) - (L^2/2mr^2)\right]\right\}^{\frac{1}{2}},$$

so

$$\int dt = \pm\int\left\{2/m\left[E_0 - V(r) - L^2/(2mr^2)\right]\right\}^{-\frac{1}{2}} dr. \tag{2.9}$$

The scattering angle can be determined as a function of r using the conservation of angular momentum, (2.7c), to give

$$\phi = \phi_0 \pm \int_{r_0}^{r} L/\left(mr^2\right)\left[2/m\left(E_0 - V(r) - L^2/2mr^2\right)\right]^{-\frac{1}{2}} dr, \tag{2.10}$$

where r_0 is the initial value of r.

It is more convenient in scattering theory to express the integral (2.10) in terms of the impact parameter p and the distance of closest approach R.

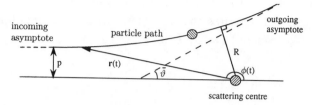

Fig. 2.1 A collision event in a central force field.

These quantities are defined in Figure 2.1, which also defines the scattering angle $\bar{\theta}$.

If the speed of the particle before collision is v_0, then

$$L = m v_0 p = (2mE_0)^{\frac{1}{2}} p. \tag{2.11}$$

This allows equation (2.10) to be rewritten

$$\phi = \phi_0 \pm p \int_{r_0}^{r} r^{-2} \left[1 - V(r)/E_0 - p^2/r^2 \right]^{-\frac{1}{2}} dr. \tag{2.12}$$

The distance of closest approach occurs when $dr/d\phi = 0$, which from equation (2.12) gives

$$1 - V(R)/E_0 - p^2/R^2 = 0. \tag{2.13}$$

From Figure 2.1 we see that initially $\phi = \pi$ and ϕ decreases as r decreases to R, so the positive square root is taken in equation (2.12). After R, ϕ continues to decrease while r increases, thus the negative root is taken after the trajectory has passed through R, and so

$$\bar{\theta} = \pi - 2p \int_{R}^{\infty} r^{-2} \left[1 - V(r)/(E_0 - p^2/r^2) \right]^{-\frac{1}{2}} dr. \tag{2.14}$$

2.1.2 Inelastic scattering

Inelastic losses are often modelled in binary collisions by the assumption of an energy loss Q, which occurs at the distance of closest approach. If $\dot{\phi}_1$ and $\dot{\phi}_2$ are the angular velocities of the particle at the instantaneous moments of time just before and just after the distance of closest approach has been reached, then from (2.8)

$$\dot{\phi}_1^2 = \frac{2}{mR} \left[E_0 - V(R) \right],$$

$$\dot{\phi}_2^2 = \frac{2}{mR} \left[E_0 - V(R) - Q \right],$$

so

$$\frac{\dot{\phi}_2}{\dot{\phi}_1} = \left(\frac{E_0 - V(R) - Q}{E_0 - V(R)}\right)^{\frac{1}{2}} = k \tag{2.15}$$

say. Thus, at the distance of closest approach, the angular momentum undergoes a step function change from its value mv_0p (given by (2.11)) to kmv_0p. The deflection angle after scattering is therefore modified. Equation (2.8) becomes

$$\dot{r}^2 = \frac{2}{m}[E_0 - V(r) - Q] - \frac{k^2 v_0^2 p^2}{r^2}, \tag{2.16}$$

$$\frac{d\phi}{dr} = \frac{\dot{\phi}}{\dot{r}} = \frac{v_0 p}{r^2}\left(\frac{2}{mk^2}(E_0 - V(r) - Q) - \frac{v_0^2 p^2}{r^2}\right)^{-\frac{1}{2}},$$

so

$$\phi = \phi(R) + p'\int_R^r r^{-2}\left[1 - V(r)/(E_0 - Q) - p'^2/r^2\right]^{-\frac{1}{2}} dr, \tag{2.17}$$

where

$$p' = \left(\frac{E_0 k^2}{E_0 - Q}\right)^{\frac{1}{2}} p.$$

If we define a reduced energy f by

$$f^2 = \frac{E_0 - Q}{E_0} \tag{2.18}$$

then the scattering angle after collision is given by a similar integral to the scattering angle before collision but with a reduced energy $f^2 E_0$ and an impact parameter $p' = (k/f)p$. Equation (4.13) gives a possible method for calculating Q.

Thus, for an inelastic binary collision, the total scattering angle is given by

$$\begin{aligned} \phi_m = \pi - p\int_R^\infty r^{-2}\left[1 - V(r)/E_0 - p^2/r^2\right]^{-\frac{1}{2}} dr \\ - p'\int_R^\infty r^{-2}\left[1 - V(r)/(E_0 - Q) - p'^2/r^2\right]^{-\frac{1}{2}} dr. \end{aligned} \tag{2.19}$$

If we define

$$\bar{\theta} = \pi - 2p\int_R^\infty r^{-2}\left[1 - V(r)/E_0 - p^2/r^2\right]^{-\frac{1}{2}} dr,$$

$$\bar{\theta}' = \pi - 2p'\int_R^\infty r^{-2}\left[1 - V(r)/(E_0 - Q) - p'^2/r^2\right]^{-\frac{1}{2}} dr.$$

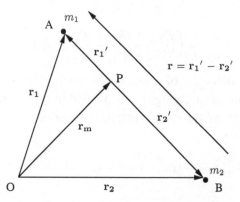

Fig. 2.2 Particles of mass m_1 and m_2 are located at points A and B whose position vectors are r_1 and r_2 from the origin 0. The centre of mass is located at P. The vectors r_1' and r_2' define the positions of m_1 and m_2 relative to P and $r = r_1' - r_2'$.

The total scattering angle is given by

$$\phi_m = \frac{1}{2}\left(\bar{\theta} + \bar{\theta}'\right). \tag{2.20}$$

The asymptotic form for large r is given by

$$\phi(r) = \phi_m + p' \int_r^\infty r^{-2}\left[1 - V(r)/(E_0 - Q) - p'^2/r^2\right]^{-\frac{1}{2}} dr$$

and, for large r, $\quad V << E_0 - Q$ and $p' << r$

$$\phi(r) \approx \phi_m + p' \int_r^\infty r^{-2}\, dr = \phi_m + p'/r. \tag{2.21}$$

2.1.3 The two-body scattering problem

Reduction to a fixed scattering centre problem. The results presented in Sections 2.1.1 and 2.1.2 were derived on the basis that the central force emanates from a point fixed in space. In two-body collisional problems, this is not the case and the force depends on the distance between two particles, each of which is free to move. The two-body problem can be reduced to an equivalent one-body problem by means of a co-ordinate transformation, which measures distances from their centre of mass. The argument is as follows.

Let the position vectors of the two particles and their centre of mass be defined as in Figure 2.2. The kinetic energy of the system is as follows:

$$\begin{aligned} T &= \frac{1}{2}m_1 \left(\dot{\mathbf{r}}_m + \dot{\mathbf{r}}_1' \right)^2 + \frac{1}{2}m_2 \left(\dot{\mathbf{r}}_m + \dot{\mathbf{r}}_2' \right)^2 \\ &= \frac{1}{2}(m_1 + m_2)\dot{\mathbf{r}}_m^2 + \frac{1}{2}m_1\dot{\mathbf{r}}_1'^2 + \frac{1}{2}m_2\dot{\mathbf{r}}_2'^2 \\ &\quad + \dot{\mathbf{r}}_m \cdot \left(m_1\dot{\mathbf{r}}_1' + m_2\dot{\mathbf{r}}_2' \right). \end{aligned} \tag{2.22}$$

However, the centre of mass is defined by

$$m_1\mathbf{r}_1' + m_2\mathbf{r}_2' = 0, \tag{2.23}$$

so

$$\begin{aligned} \dot{\mathbf{r}}_1' &= \frac{m_2}{m_1 + m_2}\mathbf{r}, \\ \mathbf{r}_2' &= -\frac{m_1}{m_1 + m_2}\mathbf{r}. \end{aligned} \tag{2.24}$$

Substituting this into (2.22) gives

$$T = \frac{1}{2}(m_1 + m_2)\dot{\mathbf{r}}_m^2 + \frac{1}{2}\frac{m_1 m_2}{m_1 + m_2}\dot{\mathbf{r}}^2. \tag{2.25}$$

For an interaction potential $V(r)$, which depends only on the particle separation, the Lagrangian $\mathcal{L}$ for the system is given by

$$\mathcal{L} = T - V(r) = \frac{1}{2}(m_1 + m_2)\dot{\mathbf{r}}_m^2 + \frac{1}{2}\frac{m_1 m_2}{m_1 + m_2}\dot{\mathbf{r}}^2 - V(r). \tag{2.26}$$

Lagrange's equations of motion in terms of the generalised co-ordinates $q = (r, r_m)$ and $\dot{q} = (\dot{r}, \dot{r}_m)$ are

$$\frac{d}{dt}\left(\frac{\partial \mathcal{L}}{\partial \dot{q}_i} \right) - \frac{\partial \mathcal{L}}{\partial q_i} = 0.$$

Since V does not depend on r_m, the equation for the centre of mass co-ordinate reduces to

$$(m_1 + m_2)\ddot{\mathbf{r}}_m = 0.$$

Thus the centre of mass is either at rest or moving uniformly with

$$\dot{\mathbf{r}}_m = \text{constant vector.} \tag{2.27}$$

The remainder of the Lagrangian (2.26) is equivalent to that due to a single particle at a distance r from a centre of force having a mass

$$\mu = \frac{m_1 m_2}{m_1 + m_2}.$$

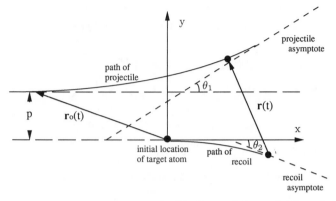

Fig. 2.3 The trajectories of the projectile (incoming particle) and the recoil.

Calculation of the path asymptotes of the scattered particles. It is usual to take a co-ordinate system that is centred at the initial position of the target atom. The co-ordinate system is defined below in Figure 2.3.

The separation between the particles is $r(t)$ and $r_0(t)$ is the separation at time $t = 0$, when the particles are sufficiently far apart that the recoil has not yet moved.

The relationship between time and the distance between the particles is given from equations (2.9) and (2.16) in the reduced mass system. After collision

$$t = \frac{1}{v_0} \int_R^{r_0} g(r)\,dr + \frac{1}{fv_0} \int_R^{r} g'(r)\,dr, \qquad (2.28)$$

where

$$g(r) = \left[1 - V(r)/E - p^2/r^2\right]^{-\frac{1}{2}},$$

$$g'(r) = \left[1 - V(r)/(E - Q) - p'^2/r^2\right]^{-\frac{1}{2}}$$

and where E is now defined in terms of the reduced mass by $E = \frac{1}{2}\mu v_0^2$.

The approach here follows closely that of Eckstein (1991). Equation (2.28) can be rewritten

$$v_0 t = \int_R^{r_0} (g - \bar{g})\,dr + \frac{1}{f}\int_R^{r} (g' - \bar{g}')\,dr$$
$$+ \int_R^{r_0} \bar{g}\,dr + \frac{1}{f}\int_R^{r} \bar{g}'\,dr, \qquad (2.29)$$

where

$$\bar{g} = \left(1 - p^2/r^2\right)^{-\frac{1}{2}},$$

$$\bar{g}' = \left(1 - p'^2/r^2\right)^{-\frac{1}{2}}.$$

Defining τ and τ' (sometimes known as time integrals) by

$$\tau = \left(R^2 - p^2\right)^{\frac{1}{2}} - \int_R^\infty (g(r) - \bar{g}(r))\,dr,$$

$$\tau' = \left(R^2 - p'^2\right)^{\frac{1}{2}} - \int_R^\infty (g'(r) - \bar{g}'(r))\,dr$$

(2.30)

we see that for large r, r_0 equation (2.29) becomes

$$v_0 t \approx -\tau + \frac{\tau'}{f} + \left(r_0^2 - p^2\right)^{\frac{1}{2}} + \left(r^2 - p'^2\right)^{\frac{1}{2}}$$

or

$$r(t) = \tau' + f\tau - fr_0 + fv_0 t.$$

(2.31)

In order to calculate the paths of the scattered and recoiled particles it is necessary to make another assumption about the collision. If the collision is elastic then linear momentum is conserved. We assume that linear momentum is also conserved when inelastic losses occur. If the co-ordinates of the incoming particle are (x_1, y_1) and those of the recoil (x_2, y_2) at time t, then conservation of linear momentum gives

$$m_1 \dot{x}_1 + m_2 \dot{x}_2 = m_1 v_0,$$

(2.32a)

$$m_1 \dot{y}_1 + m_2 \dot{y}_2 = 0.$$

(2.32b)

Equation (2.32a) integrates to give

$$x_1 + A x_2 = v_0 t + x_1(0) \approx v_0 t - r_0$$

(2.33)

and (2.32b) gives

$$y_1 + A y_2 = p,$$

(2.34)

where $A = m_2/m_1$ is the mass ratio.

Now from equations (2.21) and (2.31) for large r

$$x_1 - x_2 = r \cos \phi_m - p' \sin \phi_m,$$

(2.35)

$$y_1 - y_2 = r \sin \phi_m + p' \cos \phi_m$$

(2.36)

and solving (2.33)–(2.36) gives the asymptote path for the scattered particle

$$
y = \frac{Af \sin \phi_m}{1 + Af \cos \phi_m} x + \frac{\tau_m A \sin \phi_m}{(1 + A)(1 + Af \cos \phi_m)}
$$
$$
+ \frac{A^2 p' f + A (p' + pf) \cos \phi_m + p}{(1 + Af \cos \phi_m)(1 + A)}
\tag{2.37}
$$

and for the recoil

$$
y = \frac{-f \sin \phi_m x}{1 - f \cos \phi_m} + \frac{p'(f - \cos \phi_m) - \tau_m \sin \phi_m + p(1 - f \cos \phi_m)}{(1 + A)(1 - f \cos \phi_m)},
\tag{2.38}
$$

where $\tau_m = \tau + f\tau'$.

These two equations contain implicitly the scattering angles for the recoil and incident particle and the intercepts of the asymptotes with the axes.

2.1.4 Velocity components and energy transfer

The velocity components can be derived from the time differential of (2.31) together with (2.32). This gives for the velocities a long time after the collision

$$
\dot{x}_1 = \frac{v_0(1 + Af \cos \phi_m)}{1 + A},
$$
$$
\dot{y}_1 = \frac{Af v_0 \sin \phi_m}{1 + A},
$$
$$
\dot{x}_2 = \frac{v_0(1 - f \cos \phi_m)}{1 + A},
\tag{2.39}
$$
$$
\dot{y}_2 = -\frac{f v_0 \sin \phi_m}{1 + A}.
$$

Thus the kinetic energy of the scattered particle after collision is given by

$$
\frac{E_1}{E_0} = \frac{1}{(1 + A)^2} \left(1 + A^2 f^2 + 2Af \cos \phi_m \right)
\tag{2.40}
$$

and the recoil by

$$
\frac{T}{E_0} = \frac{A}{(1 + A)^2} \left(1 + f^2 - 2f \cos \phi_m \right).
\tag{2.41}
$$

Note that $E_0 = \frac{1}{2} m_1 v_0^2$ differs from $E = \frac{1}{2} \mu v_0^2$, the kinetic energy in the reduced mass system. The maximum transferable energy, T_m, for $f = 1$ is given from (2.41) by $\cos \phi_m = -1$ (head-on collision),

$$
T_m = \frac{4A}{(1 + A)^2} E_0 = \frac{4}{1 + A} E.
\tag{2.42}
$$

Thus (2.41) can be rewritten

$$T = \frac{1}{4} T_m \left(1 + f^2 - 2f \cos \phi_m \right), \qquad (2.43)$$

where T is the energy transferred in the collision.

Equations (2.37) and (2.38) allow us to determine the scattering angle θ_1 of the projectile and the scattering angle θ_2 of the recoil:

$$\tan \theta_1 = \frac{Af \sin \phi_m}{1 + Af \cos \phi_m}; \qquad \tan \theta_2 = \frac{-f \sin \phi_m}{1 - f \cos \phi_m}.$$

For use in ion scattering spectroscopy, it is clearly preferable to express the energy changes in terms of θ_1 and θ_2 in place of ϕ_m.

This leads to alternative forms of equations (2.40) and (2.41), namely,

$$\frac{E_1}{E_0} = \left(\frac{\cos \theta_1 \pm (A^2 f^2 - \sin^2 \theta_1)^{\frac{1}{2}}}{1 + Af} \right)^2, \qquad (2.40a)$$

$$\frac{T}{E_0} = \frac{A}{(1 + A)^2} \left[\cos \theta_2 + (f^2 - \sin^2 \theta_2)^{\frac{1}{2}} \right]^2. \qquad (2.41a)$$

In equation (2.40a), the positive sign applies for $Af > 1$ and both signs are possible for $Af \leq 1$. In the latter case the scattering angle is clearly limited by the requirement that $\theta_{\max} \leq \arcsin Af$. The accuracy with which equation (2.40a) predicts the experimentally observed peak energies in ion scattering spectroscopy is illustrated in Figure 2.4 (Hagstrum, 1954) for 2 keV He$^+$ scattering at a copper surface and 1.5 keV Ar$^+$ scattering from a gold surface. In both cases f is assumed to be 1. The scattering angle θ_1 was arranged to be 90° in the experiment, which means that equation (2.40a) simplifies to give

$$E_b = E_1/E_0 = \frac{A - 1}{A + 1}.$$

For He–Cu, $E_b = 0.88$; for Ar–Au, $E_b = 0.66$, in good agreement with the experimental results.

Note that there is a broadening ΔE_1 of the peak in the energy distribution E_1 about E_b. This broadening can be caused by isotopic or instrumental effects. In ion scattering spectroscopy it is often m_2 that is required to be determined. The mass resolution of the system $\Delta m_2/m_2$ can easily be determined in terms of $\Delta E_1/E_1$ for fixed θ_1 by direct differentiation of equation (2.40a).

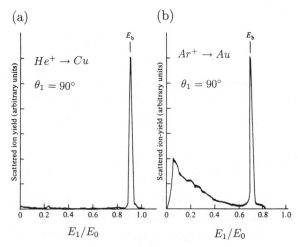

Fig. 2.4 Experimentally measured energy distributions for (a) 2 keV $He^+ \to Cu$, (b) 1.5 keV $Ar^+ \to Au$.

2.1.5 Determination of the scattering angle and the time integrals

The momentum (or impulse) approximation for elastic scattering. In this approximation, it is assumed that the initial particle path is undeflected to first order and thus the recoil receives an impulse in the direction perpendicular to this path.

The energy transferred to the recoil atom is given by $\triangle E = \frac{1}{2} m_2 v_2^2$ and the impulse I on the recoil is

$$I = \int F \, dt = m_2 v_2 = m_1 v_1,$$

where v_1 is the perpendicular velocity component of the first particle after collision.

Thus

$$\triangle E = \frac{I^2}{2m_2}.$$

I is given by $I = \int_{-\infty}^{\infty} F_\perp \, dt$, where $F_\perp$ is the component of force exerted on the stationary atom by the moving atom, perpendicular to the path of the moving atom

$$F_\perp = -V'(r) \sin \theta$$

$$I = \int_{-\infty}^{\infty} -V'(r) \sin \theta \, dt, \quad dt = \frac{dx}{v_0} = \left(\frac{m_1}{2E_0} \right)^{\frac{1}{2}} dx,$$

$$I = -\left(\frac{2m_1}{E_0} \right)^{\frac{1}{2}} p \int_0^{\infty} V'(r) \frac{dr}{(r^2 - p^2)^{\frac{1}{2}}}.$$

The deflection angle of particle 1 is a second-order effect given by arctan v_1/v_0:

$$\bar{\phi} = -\frac{p}{E} \int_p^\infty V'(r) \frac{dr}{(r^2 - p^2)^{\frac{1}{2}}}. \tag{2.44}$$

For a number of important potentials this integral can be evaluated analytically. However, it turns out to be only a reasonably good approximation for small deflection angles, $\bar{\phi} \lesssim 5°$. Equation (2.44) can also be derived by another method.

Expansion of the scattering angle and time integrals for small V/E. For elastic scattering from equation (2.19)

$$\bar{\theta} = \bar{\phi} = \pi - 2p \int_R^\infty \frac{dr}{r[r^2(1 - V/E) - p^2]^{\frac{1}{2}}}. \tag{2.45}$$

If we put $\varepsilon = 1/E$, make the substitution

$$r^2 (1 - \varepsilon V(r)) = u^2$$

and expand equation (2.45) for small ε we obtain

$$\bar{\phi} = \pi - 2p \int_p^\infty \frac{du}{u(u^2 - p^2)^{\frac{1}{2}}} \left(1 + \frac{\varepsilon u}{2} \frac{dV}{du} + O\left(\varepsilon^2\right)\right).$$

This gives to first order in ε

$$\bar{\phi} = -\frac{p}{E} \int_p^\infty V'(u) \left(u^2 - p^2\right)^{-\frac{1}{2}} du, \tag{2.46}$$

in agreement with (2.44). Unfortunately the expansion is only slowly convergent and the inclusion of further terms only improves the accuracy marginally. It is therefore better to evaluate the scattering angle by numerical quadrature or by using numerical fits such as Biersack's 'magic' formula (Biersack and Haggmark, 1980).

The time integral, τ, can be rewritten from (2.30) as

$$\tau = \left(R^2 - p^2\right)^{\frac{1}{2}} - \int_R^\infty \left(\frac{r - rV/E - r^2/(2E)}{[r^2(1 - V/E) - p^2]^{\frac{1}{2}}} - \frac{r}{(r^2 - p^2)^{\frac{1}{2}}} \right. $$
$$\left. + \frac{rV/E + r^2V'/(2E)}{[r^2(1 - V/E) - p^2]^{\frac{1}{2}}}\right) dr.$$

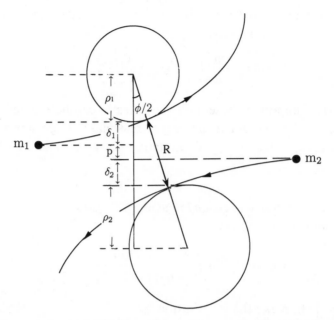

Fig. 2.5 The geometry and definition of quantities used in the 'magic' approximation.

Integrating gives

$$\tau = \left(R^2 - p^2\right)^{\frac{1}{2}} - \left\{\left[r^2\left(1 - V/E\right) - p^2\right]^{\frac{1}{2}} - \left(r^2 - p^2\right)^{\frac{1}{2}}\right\}_R^\infty$$
$$- \int_R^\infty \frac{rV/E + r^2 V'/(2E)}{\left[r^2\left(1 - V/E\right) - p^2\right]^{\frac{1}{2}}} dr. \tag{2.47}$$

For potentials such that $rV(r) \to 0$ as $r \to \infty$ we have

$$\tau = -\int_R^\infty \frac{rV/E + r^2 V'/(2E)}{\left[r^2\left(1 - V/E\right) - p^2\right]^{\frac{1}{2}}} dr.$$

If we expand for small V/E, the dominant term is given by

$$\tau = -\int_p^\infty \frac{rV/E + r^2 V'/(2E)}{\left(r^2 - p^2\right)^{\frac{1}{2}}} dr. \tag{2.48}$$

Biersack's 'magic' approximation. At the distance of closest approach, the magnitudes of the radial forces F on the two particles are equal:

$$\frac{m_1 \bar{v}_1^2}{\rho_1} = \frac{m_2 \bar{v}_2^2}{\rho_2} = -V'(R).$$

Define $\tilde{p} = p_1 + p_2 = \left(m_1 \bar{v}_1^2(R) + m_2 \bar{v}_2^2(R) \right) / F = -2E_R/V'(R)$, where E_R is the kinetic energy at the distance of closest approach

$$\tilde{p} = -2\left[E - V(R) \right] / V'(R). \tag{2.49}$$

The scattering angle is given by

$$\cos\left(\frac{\bar{\phi}}{2}\right) = \frac{p + \tilde{p} + \delta}{R + \tilde{p}}, \tag{2.50}$$

where the impact parameter p is given, R is determined from (2.13), $\tilde{p}$ from (2.49) and δ is given by a fitting formula

$$\delta = \delta_1 + \delta_2 = D(R - p)/(1 + G),$$

where

$$D = 2a\varepsilon_F \left(p/a_F \right)^b,$$

$$G = c\left[\left(1 + D^2 \right)^{\frac{1}{2}} - D \right]^{-1},$$

$$a_F = 0.8854a_B / \left(Z_1^{\frac{1}{2}} + Z_2^{\frac{1}{2}} \right)^{2/3},$$

for the Molière potential and

$$a_F = 0.8854a_B / \left(Z_1^{0.23} + Z_2^{0.23} \right)$$

for the Kr–C and ZBL potentials, where $\varepsilon_F = 4\pi\varepsilon_0 a_F / \left(Z_1 Z_2 e^2 \right)$,

$$a = 1 + C_1 \varepsilon_F^{-\frac{1}{2}},$$

$$b = \left(C_2 + \varepsilon_F^{\frac{1}{2}} \right) / \left(C_3 + \varepsilon_F^{\frac{1}{2}} \right),$$

$$c = (C_4 + \varepsilon_F) / (C_5 + \varepsilon_F),$$

In the above formula C_1–C_5 are fitting constants that depend on the potential, Z_1 and Z_2 are the atomic numbers of the impacting particle and recoil, e is the electronic charge and a_B is the Bohr radius, 5.2918×10^{-11} m. Values of C_1–C_5 for some typically used potentials are given in Table 2.1.

Gauss–Chebyshev quadrature for $\bar{\phi}$ and τ. Equation (2.20) shows that to determine $\bar{\phi}$ we need to evaluate integrals of the form

$$I = 2 \int_R^\infty r^{-2} \left[1 - V(r)/E - p^2/r^2 \right]^{-\frac{1}{2}} dr. \tag{2.51}$$

Table 2.1. *Values of constants used in the 'magic'*
formula for the scattering angle for some commonly used
interaction potentials. (See Chapter 3 for a description
of these potentials.)

	Molière	Kr $-$ C	ZBL
C_1	6.743×10^{-1}	1.0144	9.9229×10^{-1}
C_2	9.611×10^{-3}	2.3581×10^{-1}	1.1615×10^{-2}
C_3	5.175×10^{-3}	1.26×10^{-1}	7.1222×10^{-3}
C_4	6.134	6.3935×10^{-4}	9.3066
C_5	1.0×10^1	8.355×10^4	1.4813×10^1

If we rewrite by changing the variable from r to u where $u = R/r$ and define
a function

$$G(u) = \begin{cases} \left[1 - V(R/u)/E - p^2u^2/R^2\right]^{\frac{1}{2}} & 0 \le u \le 1 \\ \\ \left[1 - V(-R/u)/E - p^2u^2/R^2\right]^{\frac{1}{2}} & -1 \le u \le 0 \end{cases}$$

then

$$I = \int_{-1}^{1} \left(1 - u^2\right)^{-\frac{1}{2}} \left[\left(1 - u^2\right)^{\frac{1}{2}} G(u)\right] du. \tag{2.52}$$

The Gauss–Chebyshev m-point quadrature formula states that

$$\int_{-1}^{1} \left(1 - x^2\right)^{-\frac{1}{2}} f(x)\, dx \approx \frac{\pi}{m} \sum_{k=1}^{m} f\left(\cos \frac{(2k-1)\pi}{2m}\right), \tag{2.53}$$

so the integral in (2.45) may be rewritten

$$I = \frac{2\pi}{m} \sum_{k=1}^{m} b_k b_{m-k+1} G(a_k), \tag{2.54}$$

where

$$b_k = \cos \frac{(2k-1)\pi}{4m},$$
$$a_k = 2b_k^2 - 1.$$

The integrals in equation (2.30) can also be rewritten by changing the variable from r to $u = R/r$. If we define $H(u)$ by

$$\left[\left(I - V\left(\frac{R}{u} \right) \right) \Big/ E - \frac{p^2 u^2}{R^2} \right)^{-\frac{1}{2}} - \left(1 - \frac{p^2 u^2}{R^2} \right)^{-\frac{1}{2}} \right] \Big/ u^2 ; \ 0 \le u \le 1,$$

$$\left[\left(1 - V\left(-\frac{R}{u} \right) \right) \Big/ E - \frac{p^2 u^2}{R^2} \right)^{-\frac{1}{2}} - \left(1 - \frac{p^2 u^2}{R^2} \right)^{-\frac{1}{2}} \right] \Big/ u^2 ; -1 \le u \le 0,$$

then

$$\tau = \left(R^2 - p^2 \right)^{\frac{1}{2}} - \frac{\pi R}{m} \sum_{k=1}^{m} b_k b_{m-k+1} H\left(a_k \right). \tag{2.55}$$

For large E the second term in (2.55) is of the same order of magnitude as the first. To determine the asymptotic paths accurately requires knowledge both of $\bar{\phi}$ and of τ and so τ is often taken to be zero.

A Gauss–Chebyshev quadrature therefore provides a consistent approach but the ten-point formula for the scattering angle is approximately a factor of seven slower than the 'magic' formula which is of the same order of accuracy.

2.1.6 Cross-sections

Scattering cross-sections in a central force field. Scattering is the process by which a moving particle changes direction as a result of its interaction with a force field. So far we have examined the interaction of a single incoming particle. However, for many applications in surface science we have a beam of incoming particles, which may be treated on a continuum basis.

The scattering cross-section σ_s and the differential scattering cross-section σ_D are defined as follows

$$d\sigma_s = \sigma_D \, d\Omega = N_\Omega / N, \tag{2.56}$$

where N is the incident flux density and N_Ω the number of particles scattered into solid angle $d\Omega$ per unit time. Both σ_s and σ_D have the dimension of area. The solid angle $d\Omega$ is the element of solid angle in the direction Ω.

For a central force field there is axial symmetry so the element of solid angle can be written

$$d\Omega = 2\pi \sin \phi \, d\phi \tag{2.57}$$

where ϕ is now defined as in Figure 2.6.

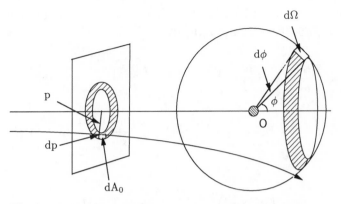

Fig. 2.6 Scattering of an incident beam by a central force.

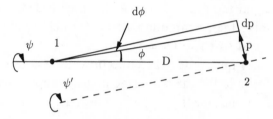

Fig. 2.7 Double scattering by a beam hitting particle 1 and then scattering a second time off particle 2.

The particles originally between p and $p + dp$ are scattered into a solid angle $d\Omega$ lying between ϕ and $\phi + d\phi$, so

$$2\pi p \, dp = -d\sigma_s = -\sigma_D \, d\Omega = -2\pi \sin \phi \sigma_D \, d\phi \qquad (2.58)$$

with the negative sign because an increase in dp leads to a decrease in $d\phi$.

Double scattering. Figure 2.6 shows that the flux through a small area dA_0 is equal to the flux through $d\Omega$

$$Np \, d p \, d\psi = N\sigma_D \sin \phi \, d\phi \, d\psi,$$

where ψ is the azimuthal angle. For a subsequent second scattering event through an angle ϕ', the flux passing into $d\Omega'$ will be

$$N'\sigma_D' \sin \phi' \, d\psi.$$

However, N' is not the same as N. If a collision has been undergone first, the flux N' is not uniform. Figure 2.7 illustrates the calculation of N'.

The flux across a surface element of the sphere of radius 1 is

$$N\sigma_D \sin \phi \, d\phi \, d\psi. \qquad (2.59)$$

We must find the flux N' across a surface element at atom 2 whose area is $p \, dp \, d\psi'$

$$N'p \, dp \, d\psi' = N\sigma_D(\phi) \sin \phi \, d\phi \, d\psi'.$$

From Figure 2.7 $dp = D \cos \phi \, d\phi$ and $d\psi = \cos \phi \, d\psi'$, thus

$$N' = \frac{\sigma_D(\phi) \sin \phi N}{D^2}.$$

The total number scattered into solid angle $d\Omega$ from a double collision will be

$$N\frac{\sigma_D(\phi) \sin \phi \sigma_D(\phi') \sin \phi'}{D^2} d\phi \, d\psi. \tag{2.60}$$

To determine the ratio of particles, Y_R, undergoing single scattering through an angle $\phi + \phi'$ to those undergoing double scattering through an angle ϕ followed by ϕ' we then divide (2.60) by (2.59) evaluated at $\phi + \phi'$. This gives

$$Y_R = \frac{\sigma_D(\phi) \sin \phi \, \sigma_D(\phi') \sin \phi'}{D^2 \sigma_D(\phi + \phi') \sin (\phi + \phi')}.$$

This formula is useful in ion scattering spectroscopy.

Scattering cross-sections in a binary collision. Equations (2.37) and (2.38) show that the incoming particle is scattered into an angle given by

$$\tan \theta_1 = \frac{Af \sin \phi}{1 + Af \cos \phi}; \qquad \cos \theta_1 = \frac{1 + Af \cos \phi}{(1 + 2Af \cos \phi + A^2 f^2)^{\frac{1}{2}}} \tag{2.61}$$

and the recoil, into an angle

$$\tan \theta_2 = \frac{-f \sin \phi}{1 - f \cos \phi}; \qquad \cos \theta_2 = \frac{1 - f \cos \phi}{(1 - 2f \cos \phi + f^2)^{\frac{1}{2}}}, \tag{2.62}$$

where ϕ is the scattering angle in the related central force field problem, given by equation (2.19). From the definition of scattering cross-section (2.56) we see that the cross-sections for the incoming particle and recoil are related to the cross-sections in the related central force field problem by

$$\sigma_D^I (\theta, \psi) \, d\Omega^I = \sigma_D (\phi, \psi) \, d\Omega,$$

$$\sigma_D^R (\theta_2, \psi) \, d\Omega^R = \sigma_D (\phi, \psi) \, d\Omega,$$

where ψ, the azimuthal angle is the same for both systems.

Now

$$\frac{d\Omega^I}{d\Omega} = \frac{\sin \theta_1 \, d\theta_1}{\sin \phi \, d\phi}; \qquad \frac{d\Omega^R}{d\Omega} = \frac{\sin \theta_2 \, d\theta_1}{\sin \phi \, d\phi},$$

so from (2.61) and (2.62)

$$
\sigma_D^I(\theta_1, \psi) = \sigma_D(\phi, \psi) \frac{\left(1 + 2Af\cos\phi + A^2 f^2\right)^{\frac{3}{2}}}{A^2 f^2 (af + \cos\theta)},
$$

$$
\sigma_D^R(\theta_2, \psi) = \sigma_D(\phi, \psi) \frac{\left(1 + f^2 - 2f\cos\phi\right)^{\frac{3}{2}}}{f^2 (f - \cos\theta)}.
$$

(2.63)

2.1.7 Potentials used in binary collision calculations

We leave a fuller discussion regarding the form of the potential until Chapter 3. In this section we define some potentials that are commonly used both in simulation programs and in analytical models. The potentials defined below are all repulsive at small particle separation, r, and zero as $r \to \infty$. We concentrate here on two types of potential, power law and screened Coulomb.

Perhaps the simplest potential is the Coulomb potential

$$
V(r) = \frac{Z_1 Z_2 e^2}{4\pi\varepsilon_0 r} = \frac{Q_1}{r},
$$

(2.64)

which is a special case of the more general power law potential

$$
V(r) = \frac{Q_s}{r^s}.
$$

(2.65)

The 'screened' Coulomb potentials considered in this section take the form

$$
V(r) = \sum_{i=1}^{4} \frac{a_i e^{-c_i r}}{r}.
$$

(2.66)

The constants a_i and c_i for a number of screened Coulomb potentials are given in Chapter 3. We derive the results now without, at this stage, specifying exact values for a_i, c_i or Q_s.

The momentum approximation for power law and screened Coulomb potentials. For the power law potential, equation (2.65), $dV(u)/du = -sQ_s u^{-s-1}$. Substituting this into equation (2.46) and changing the variables by writing $t = 1 - p^2/u^2$ gives the scattering angle ϕ for elastic scattering as

$$
\phi = \frac{sQ_s}{2Ep} \int_0^1 p^{1-s} t^{-\frac{1}{2}} (1-t)^{\frac{s-1}{2}} dt
$$

(2.67)

$$
= \frac{sQ_s}{2Ep^s} B\left(\frac{1}{2}, \frac{s+1}{2}\right) = \frac{sQ_s \sqrt{\pi} \left(s - \frac{1}{2}\right)!}{2Ep^s (s/2)!},
$$

where $B(m, n)$ is the beta function related to the factorial function by $B(m, n) = (m + 1)!(n + 1)!/(m + n + 1)!$. The scattering cross-section is determined from equation (2.56):

$$d\sigma_s = -2\pi p \frac{dp}{d\phi} d\phi = 2\pi^{-1} B_c^{2/s} \phi^{-1-2/s} d\phi, \tag{2.68}$$

where $B_c = sQ_s \sqrt{\pi} \left(s - \frac{1}{2}\right)!/[2E(s/2)!]$ and so

$$d\sigma_D = s^{-1} B_c^{2/s} \phi^{-2/s}. \tag{2.69}$$

These expressions are valid if the scattering angle ϕ is small. If we also make the assumption that the mass of the target particle is large compared with that of the incident projectile then $A \gg 1$ and equation (2.41) shows that the energy T transferred to the recoil in the collision is approximately given by

$$T \approx (E/A)\phi^2.$$

Substituting this into equation (2.68) and writing $m = 1/s$ gives

$$d\sigma_s = \pi m B_c^{2m} \left(\frac{E}{A}\right)^{-m} T^{-1-m} dT. \tag{2.70}$$

Lindhard, Nielsen and Scharff (1968) obtained an approximation for $d\sigma_s$ in terms of a single variable t in the form

$$d\sigma_s = \pi a^2 \frac{dt}{2t^{3/2}} f(t^{1/2}), \tag{2.71}$$

where

$$t = \varepsilon^2 \frac{T}{T_m}; \quad \varepsilon = \frac{4\pi\varepsilon_0 aE}{Z_1 Z_2 e^2},$$

a is the screening radius and T_m the maximum transferable energy in the collision.

Approximate forms of f which can be used in calculating ranges are given by equation (5.59) and Table (5.1). See also Sigmund (1972).

The scattering angle can also be evaluated analytically for the screened Coulomb potentials using the small scattering angle approximation. In this case V is given by equation (2.66) and substituting into equation (2.46) gives

$$\phi = \frac{p}{E} \sum_{i=1}^{4} a_i \int_p^\infty e^{-c_i r} \left(r^2 - p^2\right)^{-\frac{1}{2}} \left(\frac{c_i}{r} + \frac{1}{r^2}\right) dr$$

$$= \sum_{i=1}^{4} \frac{a_i c_i}{E} K_1 (pc_i), \tag{2.72}$$

where K_i is a Bessel function of the second kind of order i. However, the momentum approximation for the screened Coulomb potentials does not give especially useful results for the scattering cross-sections and analytical derivation of ranges and range distributions will be limited to power law potentials.

For power law potentials with $s > 1$, the momentum approximation for the time integral τ is found by substituting (2.65) into the integral (2.48) to give

$$\tau = \frac{-Q_s(s-2)\sqrt{\pi}\left(\frac{1}{2} - s/2\right)!}{4p^{3s+1}\left(1 - s/2\right)!}. \tag{2.74}$$

The corresponding value for the screened Coulomb potential is

$$\tau = \frac{1}{2E}\sum_{i=1}^{4} a_i \left[K_0\left(pc_i\right) + pc_iK_i\left(pc_i\right)\right]. \tag{2.75}$$

Table 2.2 gives a comparison between the full evaluation of the scattering angles and time integrals for the elastic collision He $\rightarrow$ Si and the approximations given by (2.72) and (2.75).

A simulation code requires a fast evaluation of τ that is valid over a wide range of values of p and E_0. Often the 'hard sphere' approximation is used i.e. $\tau = p\tan\phi_m/2$ (Eckstein, 1991). Although this is not especially accurate, the effect on the particle asymptote is small.

If a more accurate evaluation of τ is required this can be given by Chebyshev fits to the Molière and ZBL potentials (Body and Smith, 1994) described in Chapter 3. In order to obtain such a formula for τ, first write τ in terms of non-dimensionalised units. Let $\tau = a_0\bar{\tau}$, where a_0 is the screening length. Screened Coulomb potentials can be written in the form $V(r) = Ka_0/rf(r/a_0)$. Writing $\bar{p} = p/a_0$, $\bar{R} = R/a_0$, $\epsilon = E/K$ and $\tilde{V}(u) = f(u)/u$ then

$$\bar{\tau} = \left(\bar{R}^2 - \bar{p}^2\right)^{\frac{1}{2}} - \int_{\bar{R}}^{\infty}\left[(1 - \tilde{V}/\epsilon - \bar{p}^2/u^2)^{-\frac{1}{2}} - (1 - \bar{p}^2/u^2)^{-\frac{1}{2}}\right]du.$$

Now $\bar{\tau}$ is dependent on the normalised energy ϵ and impact parameter $\bar{p}$. We define a possible scheme for evaluating $\bar{\tau}$ using Chebyshev polynomials, valid in the range $10^{-3} < \epsilon < 10$, $\phi_m(p, E) > \phi_{min}$. Here ϕ_{min} is taken as $10°$. To do this define $\tilde{\epsilon} = 0.25(\log_{10}\epsilon + 3.0)$ and $\hat{\epsilon} = 2\tilde{\epsilon} - 1$. The ranges of values for $\tilde{\epsilon}$ and $\hat{\epsilon}$ are $0 \leq \tilde{\epsilon} \leq 1$ and $-1 \leq \hat{\epsilon} \leq 1$. Define $\hat{p}(\hat{\epsilon})$ by $\phi_m(\hat{p}(\hat{\epsilon}), \hat{\epsilon}) = \phi_{min}$.

Table 2.2. *A comparison between the exact values of ϕ and τ and those given by the first term in the momentum approximation ϕ_{M1}, τ_{M1} and the second term ϕ_{M2}. The definition of lattice unit is that $2LU = 5.43$ Å, the cubic cell edge in Si.*

p (lattice units)	ϕ_{exact}	ϕ_{M1}	ϕ_{M2}	τ_{exact}	τ_{M1}
100 keV					
0.01	7.92	8.04	7.87	-7.72×10^{-4}	-7.96×10^{-4}
0.05	1.10	1.11	1.10	-2.34×10^{-5}	-2.51×10^{-5}
0.1	0.36	0.36	0.36	8.39×10^{-5}	8.38×10^{-5}
0.2	0.08	0.08	0.08	7.04×10^{-5}	7.04×10^{-5}
0.3	0.03	0.03	0.03	4.32×10^{-5}	4.32×10^{-5}
1 keV					
0.01	149.53	803	-936	1.76×10^{-3}	-7.97×10^{-2}
0.05	66.37	110.7	12.59	6.31×10^{-3}	-2.51×10^{-3}
0.1	28.24	35.99	20.84	8.80×10^{-3}	8.38×10^{-3}
0.2	7.60	8.26	7.15	6.79×10^{-3}	7.04×10^{-3}
0.3	2.72	2.81	2.67	4.31×10^{-3}	4.31×10^{-3}
100 eV					
0.01	170.78	8039	-1.66019	-7.00×10^{-2}	-7.97×10^{-1}
0.05	135.10	1107	-8709	6.86×10^{-2}	-2.51×10^{-3}
0.1	96.49	360	-1154	6.45×10^{-2}	8.38×10^{-2}
0.2	45.63	8.26	-28.4	5.08×10^{-2}	7.05×10^{-2}
0.3	21.19	28.1	14	3.58×10^{-2}	4.32×10^{-2}
0.4	10.10	11.70	9.00	2.37×10^{-2}	2.64×10^{-2}
0.5	5.00	5.40	4.81	1.53×10^{-2}	1.62×10^{-2}

The Chebyshev series for $\tilde{p}$ is given by

$$\tilde{p} = 2.3\tilde{\epsilon}^{-\frac{1}{2}} \left(T_0(\hat{\epsilon})\gamma_0/2 + \sum_{n=1}^{6} T_n(\hat{\epsilon})\gamma_n \right). \tag{2.76}$$

Now let $\hat{p} = 2\bar{p}/\tilde{p} - 1$ so that the range of values for $\hat{p}$ is $-1 \le \hat{p} \le 1$. Define $\hat{\epsilon}_i$ by $\hat{\epsilon}_i = i/5 - 1$, where $i = $ integer part$[10\tilde{\epsilon}]$. Define $\lambda = (\hat{\epsilon} - \hat{\epsilon}_i)/(\hat{\epsilon}_{i+1} - \hat{\epsilon}_i)$. The formula for $\bar{\tau}$ is given by $\bar{\tau} \approx \left(\bar{R}^2 - \bar{p}^2 \right)^{\frac{1}{2}} - I_{\text{approx}}$, where

$$I_{\text{approx}} = \frac{1}{2} T_0(\hat{p}) \left[\alpha_0(\hat{\epsilon}_i)(1 - \lambda) + \alpha_0(\hat{\epsilon}_{i+1})\lambda \right]$$
$$+ \sum_{n=1}^{3} T_n(\hat{p}) \left[\alpha_n(\hat{\epsilon}_i)(1 - \lambda) + \alpha_n(\hat{\epsilon}_{i+1})\lambda \right]. \tag{2.77}$$

The binary collision

Table 2.3. (*a*) *The coefficients α_n and γ_n for the Molière potential $\gamma_0 = 1.269$, $\gamma_1 = 0.255$, $\gamma_2 = -0.188$, $\gamma_3 = -0.052$, $\gamma_4 = 0.016$, $\gamma_5 = 0.011$ and $\gamma_6 = -0.0002$.*

α_n $\dfrac{n}{i}$	0	1	2	3
0	7.292	−0.093	−0.179	−0.083
1	6.689	−0.251	−0.190	−0.057
2	6.047	−0.324	−0.150	−0.025
3	5.297	−0.323	−0.096	−0.002
4	4.387	−0.248	−0.054	0.005
5	3.384	−0.144	−0.023	0.003
6	2.451	0.078	0.002	0.000
7	1.684	−0.049	0.014	−0.001
8	1.095	−0.026	0.013	−0.002
9	0.666	−0.009	0.010	−0.003
10	0.376	0.000	0.000	−0.002

Table 2.3. (*b*) *The coefficients α_n and γ_n for the ZBL potential: $\gamma_0 = 1.162$, $\gamma_1 = 0.275$, $\gamma_2 = -0.137$, $\gamma_3 = -0.064$, $\gamma_4 = 0.001$, $\gamma_5 = 0.016$ and $\gamma_6 = 0.01$.*

α_n $\dfrac{n}{i}$	0	1	2	3
0	7.036	−0.032	−0.171	−0.073
1	6.103	−0.126	−0.157	−0.047
2	5.229	−0.173	−0.128	−0.024
3	4.407	−0.159	−0.087	−0.010
4	3.601	−0.137	−0.051	−0.002
5	2.840	−0.121	−0.021	0.001
6	2.154	−0.100	−0.000	0.001
7	1.552	−0.066	0.009	0.000
8	1.044	−0.031	0.011	−0.002
9	0.649	−0.010	0.010	−0.003
10	0.372	0.000	0.006	−0.002

The coefficients are given in Table 2.3 for the Molière and ZBL potentials. A comparison of these formulae with an *n*-point Gauss quadrature for τ has shown that there is an increase in the speed of evaluation by a factor of about $0.88n$. Since a ten-point quadrature formula is about the minimum that would normally be required, the Chebyshev fits show a considerable increase in the speed of evaluation of τ.

The maximum error for τ derived from these tables is $0.03a_0$, where a_0 is the screening length. For the range of values $10^{-3} < \epsilon < 10$, $\phi_m(p, E) < 10°$, the contribution that the term involving τ makes to the trajectory asymptotes is small. The lower limit for ϵ of 0.001 is chosen because, below this energy, the binary collision approximation is not valid. For the Molière potential $|\bar{\tau}| < 0.05$. Suppose that $1 < \epsilon < 10$, then the error in the position of the asymptotes is given by

$$|2a_0\delta\bar{\tau}\sin\phi_1/(1+A)|,$$

where ϕ_1 is the true scattering angle. This is negligible, so we may assume that $\tau \approx 0$. For the range of values $10^{-3} < \epsilon < 1$, $\phi_m(p, E) < 10°$, it is possible to show that $|\bar{\tau}_{\mathrm{approx}} - \bar{\tau}| < 0.05$ and that the contribution to the asymptote intercepts is also small and may be ignored.

2.2 Applications

2.2.1 Rutherford scattering

For the case of the Coulomb potential the equations of motion may be integrated exactly. Equation (2.7c) shows that

$$\frac{\mathrm{d}}{\mathrm{d}t} = \frac{L}{mr^2}\frac{\mathrm{d}}{\mathrm{d}\phi},$$

so that equation (2.7d) becomes

$$\frac{L}{r^2}\frac{\mathrm{d}}{\mathrm{d}\phi}\left(\frac{L}{mr^2}\frac{\mathrm{d}r}{\mathrm{d}\phi}\right) - \frac{L^2}{mr^3} = \frac{Q_1}{r^2}. \qquad (2.78)$$

Making the change of variable $r = 1/u$ gives

$$\frac{\mathrm{d}^2u}{\mathrm{d}\phi^2} + u = -\frac{mQ_1}{L^2},$$

which has the the solution

$$u = -\frac{mQ_1}{L^2} + a\cos(\phi - \phi'),$$

where a and ϕ' are constants of integration. Thus

$$\frac{1}{r} = \frac{mk}{L^2}[1 + b\cos(\phi - \phi')],$$

where $b = -aL^2/(mQ_1)$.

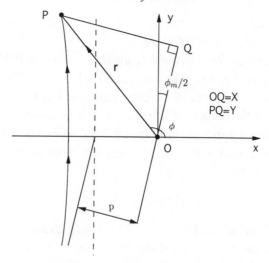

Fig. 2.8 The hyperbolic path of a particle in a Coulomb field.

Evaluating the scattering integral (2.10) gives

$$\phi = \phi' - \arccos \frac{L^2 u + mQ_1}{\left(m^2 Q_1^2 + 2E_0 mL^2\right)^{\frac{1}{2}}}, \tag{2.79}$$

$$\frac{1}{r} = -\frac{mQ_1}{L^2} \left\{ 1 + \left[1 + 2E_0 L^2 / \left(mQ_1^2\right)\right]^{\frac{1}{2}} \cos\left(\phi - \phi'\right) \right\}. \tag{2.80}$$

This equation is simplified by rotating the co-ordinates so that $\phi' = 0$. It is then of the form

$$\frac{1}{r} = c\left(1 + d \cos \phi\right), \tag{2.81}$$

where

$$c = -Q_1/(2E_0 p^2); \quad d = \left(1 + 4E_0^2 p^2 / Q_1^2\right)^{\frac{1}{2}} \tag{2.82}$$

using (2.11) to write L in terms of p.

The equation of the curve is a hyperbola but the scattering angle is such that $\cos \phi < -1/d$. The geometry of the collision is shown in Figure 2.8. The angles of the asymptote to the projectile path are such that $\cos \phi = -1/d$:

$$\cos\left(\frac{\pi}{2} + \frac{\phi_m}{2}\right) = -\frac{1}{d}$$

so

$$\sin\left(\frac{\phi_m}{2}\right) = \frac{1}{d}; \quad \cot^2\left(\frac{\phi_m}{2}\right) = d^2 - 1$$

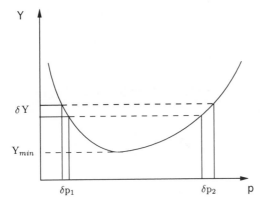

Fig. 2.9 Y as a function of p for fixed X.

and so from (2.82)

$$p = \frac{Q_1}{2E_0} \cot \left(\frac{\phi_m}{2} \right) \tag{2.83}$$

and the cross-section σ_D is given from (2.58) as

$$\sigma_D = \frac{Q_1^2}{16E_0^2} \mathrm{cosec}^2 \left(\frac{\phi_m}{2} \right). \tag{2.84}$$

One of the interesting quantities in ion scattering spectroscopy is the distance PQ defined in Figure 2.8.

$$Y = \mathrm{PQ} = r \sin \left(\phi + \frac{\phi_m}{2} - \frac{\pi}{2} \right). \tag{2.85}$$

If the angle ϕ_m is small, then expanding (2.85) for large d gives

$$Y = p + \frac{Q_1 X}{E_0 p}, \tag{2.86}$$

where $X = \mathrm{OQ}$. The distance Y has a minimum value as a function of p, for fixed X, given by

$$Y_{\min}^2 = \frac{4Q_1}{E_0} X. \tag{2.87}$$

This is a parabola and defines the edge of a wake region behind the struck particle into which the projectile cannot penetrate. This wake region is known as a *shadow cone*. For a beam of incoming particles with a constant flux and uniform direction, there is a concentration of flux at the edge of the shadow cone. The flux distribution along the line PQ defined in Figure 2.8 can also be calculated. A plot of Y as a function of p for fixed X is shown in Figure 2.9.

2.2.2 Surface spectroscopy using shadow cones

Structure. It can be seen that p is not a single-valued function of Y. From Figure 2.9 the flux $f(Y)$ through an annular loop of width δY is given by

$$f(Y)2\pi Y \, \mathrm{d}Y = 2\pi (p_1 \, \mathrm{d}p_1 + p_2 \, \mathrm{d}p_2),$$

so

$$f(Y) = \frac{p_1}{Y}\frac{\mathrm{d}p_1}{\mathrm{d}Y} + \frac{p_2}{Y}\frac{\mathrm{d}p_2}{\mathrm{d}Y}.$$

Using (2.86) we obtain

$$f(Y) = \frac{1}{2}\left[\left(1 - \frac{Y_{min}^2}{Y^2}\right)^{\frac{1}{2}} + \left(1 - \frac{Y_{min}^2}{Y^2}\right)^{-\frac{1}{2}}\right], \tag{2.88}$$

i.e. the flux is infinite at the edge of the shadow cone. This enhanced flux near the edge of the shadow cone is important in ion scattering spectroscopy (ISS) and impact collision secondary ion mass spectroscopy (ICSIMS). In ISS a beam is incident on a crystal which is rotated and the flux of back-scattered ions is monitored. When the edge of the shadow cone lines up with a second atom in the crystal we expect to see an increase in the back-scattered ion yield. From a knowledge of the width of the shadow cone and the peaks in the back-scattered spectrum we can determine by triangulation the distance between atoms in the crystal. The same principle is used in ICSIMS except that the flux of sputtered particles is measured.

The principles of the shadow cone method for surface structure determination are illustrated in Figure 2.10. In (a) the angle ψ_1 is such that the shadow cone edge lies just above particle B. At this angle ψ_1 we expect to see an increase in the scattered ion yield. In (b) the angle ψ_2 is such that the end of the shadow cone lies just below particle B. At this angle we expect to see an increase in the ejected atom yield. In (c), the edge of the shadow cone intersects particle B. At this angle we expect to see an increase in the 180° back-scattered ion yield. Empirical formulae for the critical angle ψ_c have been given by Fauster (1988) by numerical fitting for the Molière and ZBL potentials; they are

Molière

$$\ln\psi_c = 4.6239 + \ln(d/a_0)(-0.0403 \ln A_c - 0.6730)$$
$$+ \ln A_c(-0.0158 \ln A_c + 0.4647) \tag{2.89}$$

ZBL

$$\ln\psi_c = 4.7334 + \ln(d/a_0)(-0.0250 \ln A_c - 0.7205)$$
$$+ \ln A_c(-0.0094 \ln A_c + 0.3647) \tag{2.90}$$

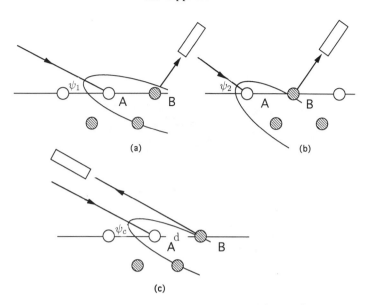

Fig. 2.10 (a)–(c) Principles of the shadow cone method for surface structure analysis.

where a_0 is the screening length and $A_c = Q_1/(Ea_0)$. These expressions are especially useful for spectroscoptists because they give direct expressions for the 180° back-scattered flux enhancement as a function of the particle spacing d. Measuring ψ_1 enables d to be determined.

An example of a shadow cone calculated directly from the trajectories of an impacting 1 keV He atom incident on a Si atom is shown in Figure 2.11. The Si atom is depicted by the black circle at the head of the cone. The other circles indicate the relative positions of the nearby atoms in the Si lattice and are shown to indicate the relative length scale of the shadow cone with regard to a crystal. The surface is Si{110}.

Composition. We may also determine the composition of a target struck by an incoming beam of particles by collecting those particles scattered at a known angle and measuring their energy. If the particles undergo only one collision before scattering, then, knowing E_0, m_1 and ϕ_m, we can determine m_2, the mass of the target atoms, by using equations (2.40) and (2.40a). The basis of the quantitative nature of this technique is that the cross-section for scattering of the beam is accurately known. This process is known as Rutherford back-scattering and is an important technique used for compositional measurements in surface analysis.

It is also possible to determine m_2 by recoil spectroscopy. The incident beam is at near grazing incidence and in this case it is the energy of the

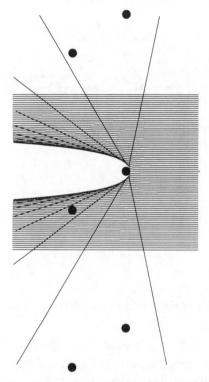

Fig. 2.11 The shadow cone for 1 keV He→Si.

struck particle at a given angle which is measured. The mass m_2 is then determined from equation (2.41). Recoil analysis can be aided by placing a foil in front of the detector to absorb the forward scattering beam particles whilst allowing through lighter recoils. Recoil analysis done this way is restricted to light elements, precisely those which are most difficult to detect by direct back-scattering methods.

2.2.3 *Projectile stopping*

If, instead of a single collision between a projectile and a target atom, we specify that the target atoms are distributed at random, with a density N (atoms/m^3), then the probability dP of a projectile to undergo a collision specified by a cross-section $d\sigma_s$ in moving through a small thickness δx of the target, is given by

$$dP = N \, \delta x \, d\sigma_s. \tag{2.91}$$

Now $d\sigma_s$ depends on the initial energy E of the projectile and also on the recoil energy T. Since T is also the energy lost by the projectile in the collision, the average energy loss δE is given by

$$\delta E = \int T \, \mathrm{d}P = -N \, \delta x \int_{T=0}^{T=T_m} T \, \mathrm{d}\sigma, \tag{2.92}$$

where T_m is the maximum transferred energy.

Equation (2.92) can be written in an alternative form as

$$\frac{\mathrm{d}E}{\mathrm{d}x} = -N S_n(E), \tag{2.93}$$

where

$$S_n(E) = \int_{T=0}^{T=T_m} T \, \mathrm{d}\sigma \tag{2.94}$$

is called the *stopping cross-section*. The quantity $\mathrm{d}E/\mathrm{d}x$ is called the stopping power.

The stopping cross-section can be easily calculated for power law potentials by substituting (2.70) into (2.94):

$$S_n(E) = \frac{m}{1-m} \pi (B_c^2 A)^m \left(\frac{4}{1+A}\right)^{1-m} E^{1-2m}. \tag{2.95}$$

Moving particles (ions) can also lose energy due to interaction with electrons. The mechanism of energy loss is discussed in more detail in Chapter 4. For the purposes of this chapter it is sufficient to note that it is possible to define an electronic stopping power S_e, equivalent to S_n, by

$$\left(\frac{\mathrm{d}E}{\mathrm{d}x}\right)_e = -N S_e(E), \tag{2.96}$$

where $(\mathrm{d}E/\mathrm{d}x)_e$ is the inelastic energy loss by the projectile due to collisions with electrons. If nuclear and electronic stopping are assumed independent then we can write

$$\frac{\mathrm{d}E}{\mathrm{d}x} = -N(S_n(E) + S_e(E)). \tag{2.97}$$

Inverting this expression, the path length travelled by a particle of initial energy E before coming to rest is

$$R(E) = \int^E \frac{\mathrm{d}E}{N(S_n(E) + S_e(E))}.$$

Ion ranges are important in technological applications such as ion implantation and are discussed in more detail in Chapters 5 and 6.

2.3 Conclusion

Although the assumption that all atomic collisions in solids can be modelled by unconnected binary collisions is clearly false, the formulae can be succesfully used in many cases. For example, the simple formula connecting the energy loss with the scattering angles for a simple two-body collision forms the successful basis of much of ion scattering spectroscopy. The ranges of high-energy particles in amorphous solids are also accurately described. The two-body results described in this chapter form the basis for much of the rest of this book. In Chapter 5 a collision cascade is described statistically in terms of Boltzmann transport theory assuming that all the collisions occur in a binary nature. In Chapter 7 algorithms based on the binary collision approximation are described together with some discussion on the validity of the approximation. The results of Section 2.2.3 on projectile stopping are extended in Chapter 6, where a comprehensive approach to the calculation of these ranges is described in detail. The key to successful simulation is to know when the assumptions used in the model are valid and when they break down. This requires a value judgement and therefore can never be entirely objective. The sometimes differing results between calculations based on the BC approximation and those using a full molecular dynamics calculation have often been the subject of lively debate at international meetings. It is the authors' view that it is important not to become too involved with one's own preferred technique but to recognise instead the strengths and weaknesses of all approaches. Differing approaches to a problem enrich and advance the knowledge of the subject.

References

Biersack, J. P. and Haggmark, L. G. (1980). *Nucl. Instrum. Meth.* **174** 257.

Body, G. and Smith, R. (1994). *Nucl. Instrum. Meth.* B **84** 425.

Eckstein, W. D. (1991). *Computer Simulation of Ion–Solid Interactions*, Springer-Verlag, Berlin.

Fauster, T. (1988). *Vacuum* **38** 129.

Hagstrum, H. D. (1954). *Phys. Rev.* **96** 336.

Lindhard, J., Nielsen, V. and Scharff, M. (1968). *Mat. Fys. Medd. Dan. Vid. Selsk.* **36** 10.

Sigmund, P. (1972). *Rev. Roum. Phys.* **17** 823.

3
Interatomic potentials

3.1 General principles

The forces between atoms lie at the very heart of almost all physical and chemical phenomena. Examples might be the material state of a substance, the structure and strength of a solid, the viscosity of a liquid or the pressure of a gas. One of the fundamental problems of science concerns the determination of these forces since they can be used to determine so many other scientific phenomena. In this book we will assume that these forces can be derived from an interatomic potential function V, which depends on the position of the atoms, so that for a system of N_A atoms whose position vectors are r_i, $i = 1, 2 \ldots N_A$, the force F_i on the ith atom is given by

$$F_i = -\frac{\partial V(r_1, r_2 \ldots r_{N_A})}{\partial r_i}.$$

As it stands, this expression for the force is genuinely a many-body one. For the case of two interacting particles only, $F = -\partial V(r)/\partial r$, where F is the force acting on either of the particles whose mutual separation is r.

The standard model of an atom is that it is constituted from a central nucleus whose diameter is of the order of 10^{-4} Å surrounded by orbital electrons, the diameter of the atom being of the order of 1 Å. In the case of two-body interactions it is possible to draw some general conclusions regarding the form of V. We would expect that $V \to 0$ as $r \to \infty$. If the interacting particles are two charged ions, the speed of decay can be estimated because, at large separation, we would expect that $V \sim 1/r$. If the particles behave like dipoles at large separation then $V \sim 1/r^6$. At very close separation we would expect strong nuclear repulsion to exist between two atoms so that $V \to \infty$ as $r \to 0$. If there is bonding between the atoms then V would have a minimum value at some intermediate distance. This is usually of the order of 2 Å. It is thus possible to sketch a curve for V

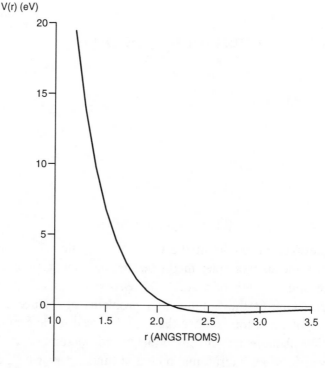

Fig. 3.1 A simple two-body interatomic potential. This is the Cu/Cu Morse potential corresponding to the second set of parameters in Table 3.5.

as a function of r which contains these attributes. Such a curve is given in Figure 3.1.

It might be expected on the basis of a knowledge of electrostatics that, for N_A interacting atoms, the potential $V(r_1, r_2 \ldots r_{N_A})$ could be written as a simple sum of these pairwise interacting potentials, since the electrostatic potentials satisfy Laplace's equation, which is linear. The interatomic potentials arise from a solution to Schrödinger's equation which is highly non-linear. Thus, in general, these potentials are not pairwise additive although pairwise addition is an assumption that is often used. We can illustrate the inapplicability of pairwise addition by an example. The binding energy of the Si dimer is 3.21 eV. Now, crystalline Si takes the form of a diamond lattice, which is tetrahedrally bonded. On the basis of pairwise addition and assuming contributions to the crystal binding energy from the nearest neighbours only, we would expect a binding energy of 6.42 eV. However, crystalline Si has a binding energy of 4.63 eV. Clearly a Si potential based on pairwise addition cannot describe accurately both the cohesive energy of the dimer and that of the lattice. It is also the case that a diamond lattice is unstable to small perturbations if the binding is modelled by pairwise additive central potentials.

The interatomic potential functions should ideally be able to describe the bonding within a crystal as well as the two-body interaction properties. This bonding depends on the atomic constituents of the crystal. For ionic crystals, we might define a model which consists of representing the atoms by oppositely charged billiard balls. For covalent materials the ultimate basis for the bonding is electrostatic but there is a strong directional bond. For metals there is a weaker bond, which, in the Sommerfeld model, is formed by unsaturated electrons moving freely through a lattice of atoms. For molecular crystals there are weak Van der Waals bonding forces. All of these physically different cases must be modelled by interatomic potential functions before dynamical calculations can be carried out.

Keeping the above principles of bonding in mind, we will describe and give parameters for some interatomic potentials which fit realistic functional forms to the potential surfaces in the region where the particle separation exceeds ≈ 1 Å. Although these mathematical functions will be based on the underlying physics of each situation, at the end of the day we are often only involved in a numerical fit to a functional form which contains free parameters, so the precise form of the many-body potentials need not be specified too exactly. Indeed, the same functional form can often be used to model different types of bonding, albeit with different fitting parameters.

We will also derive, from first principles, a useful two-body repulsive potential, the screened Coulomb potential. A functional form which approximates this screened Coulomb potential will then be used to fit experimental data. This potential will be used to describe interactions for close particle separation (i.e. $\lesssim 1$ Å).

3.1.1 Ab initio *potentials*

For a single particle such as an electron of mass m moving in a region of space under the influence of a potential $\mathcal{U}$, the particle can be described by a wavefunction $\Psi(r, t)$, which satisfies Schrödinger's equation

$$\left(-\frac{\hbar^2}{2m} \nabla^2 + \mathcal{U} \right) \Psi = i\hbar \frac{\partial \Psi}{\partial t}. \tag{3.1}$$

For a single electron orbiting a nucleus of atomic number Z, $\mathcal{U}$ would take the form

$$\mathcal{U}(r) = \frac{Ze^2}{4\pi\epsilon_0 r}$$

where r is the radial distance of the electron from the nucleus. In the case in which the potential $\mathscr{U}$ is independent of time, the variables can be separated by writing

$$\Psi(r,t) = \psi(r)\exp(-i\mathscr{E}t/\hbar) \tag{3.2}$$

where $\mathscr{E}$ is a separation constant. Substituting (3.2) into (3.1) gives the time-independent Schrödinger equation

$$\left(-\frac{\hbar^2}{2m}\nabla^2 + \mathscr{U}\right)\psi = \mathscr{E}\psi \tag{3.3}$$

or

$$\mathscr{H}\psi = \mathscr{E}\psi \tag{3.4}$$

where $\mathscr{H}$ is the Hamiltonian operator, $\mathscr{H} = -[\hbar^2/(2m)]\nabla^2 + \mathscr{U}$. The separation constant $\mathscr{E}$ has the dimension of potential. Equation (3.4) must be solved for ψ and $\mathscr{E}$ and an analytical solution is possible for the isolated electron, which is given in any standard book on quantum chemistry. The result is a set of wavefunctions known as atomic orbitals.

Most problems in chemistry and materials science involve large numbers of particles. In the general case we might have N_N nuclei and N_e electrons. If their position vectors are R_i, $i = 1,2\ldots N_N$ and r_j, $j = 1,2\ldots N_e$ respectively, then the Hamiltonian for the system is given by

$$\mathscr{H} = -\sum_{i=1}^{N_N}\frac{\hbar^2}{2M_i}\frac{\partial^2}{\partial R_i^2} - \sum_{j=1}^{N_e}\frac{\hbar^2}{2m}\frac{\partial^2}{\partial r_j^2} + \mathscr{U}_{ee} + \mathscr{U}_{Ne} + \mathscr{U}_{NN}. \tag{3.5}$$

The nuclear and electron masses are M_i and m respectively and $\mathscr{U}_{ee}$, $\mathscr{U}_{Ne}$ and $\mathscr{U}_{NN}$ are given by

$$\mathscr{U}_{ee} = \frac{1}{2}\sum_{i=1}^{N_e}\sum_{j=1}^{N_e}\frac{e^2}{4\pi\epsilon_0 \mid r_i - r_j \mid}, \quad i \neq j,$$

$$\mathscr{U}_{NN} = \frac{1}{2}\sum_{i=1}^{N_N}\sum_{j=1}^{N_N}\frac{Z_i Z_j e^2}{4\pi\epsilon_0 \mid R_i - R_j \mid}, \quad i \neq j,$$

$$\mathscr{U}_{Ne} = \sum_{i=1}^{N_e}\sum_{j=1}^{N_N}\frac{Z_i e^2}{4\pi\epsilon_0 \mid R_i - r_j \mid}.$$

Here Z_i is the atomic number of the ith nucleus. The interatomic potential is derived by solving equation (3.4) with the Hamiltonian given by (3.5). For a dynamical system of N_N atoms it is, at least in principle, possible to determine the motion by solving (3.5) for the energy $\mathscr{E}$ which determines the forces between the atoms, integrating the system for a small timestep δt, recalculating $\mathscr{E}$ with the new positions and so on. Such a procedure is extremely expensive in terms of computing time and at best can handle a system containing only a few atoms. An *ab initio* molecular dynamics scheme based on a slightly different Hamiltonian approach was derived by Car and Parrinello (1985) and applied to an eight-atom Si system. Today the motion of a few hundred atoms can be calculated using this technique. We will briefly mention some general principles concerning *ab initio* potential calculations but concentrate on describing semi-empirical formulae that approximately fit the potential surfaces. We then incorporate these formulae into computer programs that determine the motion of the dynamical system.

Computational quantum chemistry has progressed in leaps and bounds over recent years and many of the early approximations which were used to obtain solutions to Schrödinger's equation now need no longer be made. However, much of the available computational software does make use of two important simplifications. The first is the Born–Oppenheimer approximation which effectively states that the motions of the electrons and of the nuclei can be treated separately. This precludes, for example, phenomena such as electron–phonon coupling. However, for a large number of chemical phenomena the Born–Oppenheimer approximation is a useful one. It is also in line with our later approaches for modelling dynamical systems of particles in which, in many cases, we consider only the effects of momentum transfer via nuclear collisions. Electronic energy loss is treated as a separate phenomenon. A second approximation that is helpful in determining solutions to Schrödinger's equation is the self-consistent field (SCF) approximation. The basic physical idea of the SCF method is that each electron moves in an average field due to the nuclei and the remaining electrons. This assumption forms the basis of the Hartree–Fock method of solving Schrödinger's equation. The Hartree–Fock method is used in most quantum chemistry molecular electronic structure calculations. Another modern approach, which allows for *ab initio* molecular dynamics calculations, is the bond order potentials method developed at Oxford (Sutton, 1992). This contains more parameters than the approach by Car and Parrinello but requires less computing time.

There are several sophisticated public domain quantum chemistry software packages for carrying out calculations to determine the potentials, and perhaps the most well known of these is the GAUSSIAN XY series where XY stands for the year of issue. The programs are based on the work of Hehre *et al.* (1970). These packages consist of programs to calculate one- and two-electron integrals over Gaussian orbitals and to perform SCF calculations for closed and open shell molecules. They can therefore be used with care to calculate interatomic potentials for a many-body system. This is still not an especially straightforward process. The calculated potentials depend critically on the basis functions used in the method. There are many texts available in the literature that describe procedures for *ab initio* potential calculations and this is a vast subject. Thus, although we might refer to potentials calculated using quantum chemistry computer packages, we do not intend to describe these calculations here.

3.2 The repulsive wall potential

At close separation i.e. $\lesssim 1$ Å, strong repulsive forces usually exist between two particles. In a collision cascade of energetic particles, the relative velocity between two particles can be sufficient to overcome the energy barrier and a hard collision results. If the interaction potential is incorrectly given then the scattering which takes place during the collision will be wrongly calculated. It is thus important to have a reasonable model of the repulsive part of the interaction potential that is valid at small particle separation. Here a model for the potential of an isolated atom is derived from first principles and modified for the case of two particles. Many-body repulsive potentials are not discussed.

3.2.1 The screened Coulomb potential for the isolated atom

The screened Coulomb potentials are analytic formulae used to describe the potential distribution within the region of the atom where the electrons are distributed (i.e. < 1 Å). They are based on the assumption that outside the nucleus the quantum model of an electron gas can be used. In this model the independent electron approximation is invoked so that the ground state of an N_e electron system can be found by first determining the energy levels of a single electron in a volume $\mathscr{V}_L$ and then filling the levels. The appropriate solution of Schrödinger's equation for the electron gas is $\psi_k = (1/\sqrt{\mathscr{V}_L}) \exp(i\mathbf{k} \cdot \mathbf{r})$ with energy $\mathscr{E}(\mathbf{k}) = \hbar^2 k^2/(2m)$ and momentum $\mathbf{p} = \hbar\mathbf{k}$ (Ashcroft and Mermin, 1976). The normalisation constant $\mathscr{V}_L^{-\frac{1}{2}}$ is chosen so that the probability of finding an electron somewhere in the volume $\mathscr{V}_L$ is unity. The wave number $\mathbf{k}$ of an electron in the level ψ_k is quantised so that the number of allowed $\mathbf{k}$ values per unit volume of $\mathbf{k}$ space is $\mathscr{V}_L/(8\pi^3)$. For large N_e the volume of $\mathbf{k}$ space filled by the occupied levels will be approximately spherical. The radius of this sphere is denoted by k_F and thus the number of allowed values of k within the sphere is

$$\frac{4\pi k_F^3}{3} \frac{\mathscr{V}_L}{8\pi^3} = \frac{k_F^3}{6\pi^2} \mathscr{V}_L.$$

Since each k state can contain two electrons of opposed spin, in order to accommodate N_V electrons we must have

$$N_V = 2\frac{k_F^3}{6\pi^2} \mathscr{V}_L .$$

The electron density $n = N_V/\mathscr{V}_L = k_F^3/(3\pi^2)$ and the ground state is found by occupying all the single electron levels with $k < k_F$ and leaving all those with $k > k_F$ unoccupied. Now, let the potential energy for the electron be $U(r)$ and let the maximum energy of the electrons be U_0, then the kinetic energy of the electron is $\hbar^2 k_F^2/(2m)$ and

$$\frac{\hbar^2 k_F^2}{2m} - eU(r) = eU_0 .$$

So for $U > U_0$

$$n(r) = \frac{1}{3\pi^2 \hbar^3} [2me\,(U(r) - U_0)]^{\frac{3}{2}} .$$

The charge density $\rho(r) = n(r)e$ and the potential $U(r)$ can then be found by inserting this value of ρ into the radially symmetric Poisson equation

$$\frac{1}{r^2} \frac{d}{dr} \left(r^2 \frac{dU}{dr} \right) = \frac{\rho}{\epsilon_0} = \frac{n(r)e}{\epsilon_0}. \tag{3.6}$$

This equation can be rendered non-dimensional by writing

$$V(r) = U(r) - U_0 = \frac{Ze}{4\pi\epsilon_0 r} \chi(x), \tag{3.7}$$

$$r = \frac{1}{2} \left(\frac{3\pi}{4} \right)^{\frac{1}{3}} a_B Z^{-\frac{1}{3}} x = 0.8853 a_B Z^{-\frac{1}{3}} x = a_0 x,$$

where a_B is the Bohr radius and Z the atomic number of the atom. This gives the Thomas–Fermi equation

$$x^{\frac{1}{2}} \frac{d^2\chi}{dx^2} = \chi^{\frac{3}{2}}. \tag{3.8}$$

The boundary conditions for χ are obtained by assuming that the potential behaves like a simple Coulomb potential as $r \to 0$, so $\chi(0) = 1$ and also that the potential tends to zero at ∞ faster than a Coulomb potential so that $\chi \to 0$ as $x \to \infty$. This produces a boundary value problem that is not straightforward to solve numerically and some analytical work is required. The best way to solve the equation is first to obtain an asymptotic expansion for large x of the form

$$\chi(x) = \frac{A}{x^3} \sum_{k=0}^{\infty} c_k (Bx^\lambda)^k \qquad (3.9)$$

and then integrate backwards.

It is readily shown by direct substitution that for consistency $A = 144$, $c_0 = -c_1 = 1$ and $\lambda = \frac{1}{2}(7 - \sqrt{73})$. The other constants c_k are determined by a recurrence relation. The first 30 values of c_k are given by Torrens (1972). The numerical integration of (3.8) is split into two parts. First, make the substitution $x = 1/u$ and integrate from $u = \epsilon$ to $u = 1$, where the values of χ and χ' are determined at $u = \epsilon$ from the asymptotic formula assuming a value for B and that ε is some small value. Then, integrate from $x = 1$ to $x = 0$, matching the values of χ and χ' when $u = x = 1$. This in general will not give the correct value for $\chi(0)$. The value of B is adjusted until the correct value of $\chi(0)$ is obtained. The correct value of B is given by $B = 13.27097$. The numerical solution is plotted for $0 \le x \le 10$ in Figure 3.2. Outside this region, equivalent to a distance of ≈ 1 Å, depending on Z, the statistical model is invalid.

The numerical solution is cumbersome to use in computer simulation and various analytical approximations have been derived. One of the most popular is the Molière potential χ_M,

$$\chi_M(x) = 0.35e^{-0.3x} + 0.55e^{-1.2x} + 0.10e^{-6.0x}. \qquad (3.10)$$

This agrees well with the numerical solution over the range $0 \le x \le 10$, the curve being virtually indistinguishable from the Molière potential plotted in Figure 3.2 (but $\chi_M \to 0$ as $x \to \infty$ faster than the numerical solution). The other potentials shown in Figure 3.2 are defined in the next section.

3.2.2 Two-body screened Coulomb potentials

The mathematical detail involved in the derivation of an analytical form for the two-body screened Coulomb potential is quite complex and beyond the scope of this book. The paper by Firsov (1957) gives a variational derivation. A numerical fit to the solution by Firsov that retains the same analytical form for the screening function as for the isolated atom is given by

$$V = \frac{Z_1 Z_2 e^2}{4\pi\epsilon_0 r} \chi(r/a_0),$$

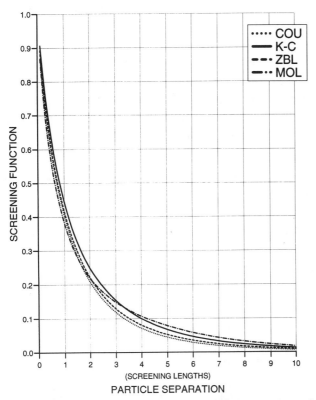

Fig. 3.2 The screening function χ for the Lenz–Jenson (L–J), Molière (MOL), ZBL and krypton–carbon (K–C) potentials. Over this region, the numerical solution to equation (3.8) is indistinguishable from the Molière potential.

where Z_1 and Z_2 are the atomic numbers of the two atoms and a_0 is the screening length defined by

$$a_0 = 0.8853 a_B \overline{Z}^{-\frac{1}{3}}. \tag{3.11}$$

The quantity $\overline{Z}$ is an 'average' atomic number of the two particles. The form favoured by Firsov is

$$\overline{Z}^{-\frac{1}{3}} = (Z_1^{\frac{1}{2}} + Z_2^{\frac{1}{2}})^{-\frac{2}{3}}$$

although Bohr (1948) used

$$\overline{Z}^{-\frac{1}{3}} = (Z_1^{\frac{2}{3}} + Z_2^{\frac{2}{3}})^{-\frac{1}{2}}.$$

These expressions for $\overline{Z}$ differ by at most a factor of $2^{\frac{1}{6}} = 1.12$ and so can be regarded as equivalent within the approximation of the Thomas–Fermi approach.

The main drawback of these two-body potentials is their relatively slow decay as $r \rightarrow \infty$. The Molière potential χ_M given in equation (3.10) has the advantage of dropping off to zero more quickly as $r \rightarrow \infty$ and can be said to represent the close-shell repulsion at intermediate values of r more accurately than does the full numerical solution for χ. Inevitably the approximations used mean that there are inaccuracies compared with a full numerical solution of Schrödinger's equation and so a_0 is often regarded as an adjustable parameter for each two-body combination, which can be matched either to SCF calculations or to experimental data.

We have thus established some general principles which form the basis by which interatomic potential functions can be described. The rest of this chapter will be concerned with describing some standard semi-empirical interatomic potential functions which are often used in computer simulation codes and in defining methods by which such potential functions can be fitted to *ab initio* calculations or experimental data.

Some of those potentials most often used in simulation codes are given below. The Molière, krypton–carbon and ZBL potentials take the form

$$\chi = \sum_{i=1}^{4} a_i \exp(-b_i x). \tag{3.12a}$$

The Lenz–Jenson potential is written

$$\chi = \exp(-q)\left(1 + \sum_{i=1}^{4} a_i q^i\right); \quad q = 3.111\,26x^{1/2}. \tag{3.12b}$$

The numerical coefficients are given in Table 3.1. Many other analytic approximations for χ have also been suggested. The reader is referred to Torren's book for a compendium and for original references. Some of the screening functions are compared in Figure 3.2. For atomic collision studies, the ZBL potential is the one that is most frequently used. This is because it has been found to be more closely in agreement with experiment (Ziegler, Biersack and Littmark 1985) than the others. The ZBL potential is used extensively in the binary collision codes such as TRIM (Biersack and Haggmark, 1980) and some simple analytic formulae have been developed to speed up the calculation of the scattering integrals, see Chapter 2. Figure 3.3 compares the ZBL potential for Si–Cu interactions with self-consistent field calculations. The agreement is reasonable but not perfect.

Another form for the repulsive wall potential that is often used is the Born–Mayer potential. This takes the form

$$V(r) = A \exp(-br). \tag{3.13}$$

The Born–Mayer potential is less satisfactory to use in atomic collision

Table 3.1. *Numerical coefficients for screened Coulomb potentials*

	a_1	a_2	a_3	a_4
Molière	0.35	0.55	0.1	0
Krypton–carbon	0.19095	0.47367	0.33538	0
ZBL	0.1818	0.5099	0.2802	0.02813
Lenz–Jenson	1.0	0.3344	0.0486	0.002647

	b_1	b_2	b_3	b_4
Molière	0.3	1.2	6.0	0
Krypton–carbon	0.27854	0.63717	1.91926	0
ZBL	3.2	0.9423	0.4029	0.2016

studies because the values of A and b must be determined independently for each pair combination and also $V(0) = A \neq \infty$. A table of values for A and b is given by Torrens (1972). This form of potential is used to model the near-neighbour repulsion for certain ionic materials described in Section 3.3.

3.3 The attractive well potential

For a two-particle system and particle separations of around 1-4 Å, the interatomic potential function is often attractive. Huber and Herzberg (1979) have compiled a compendium of experimental data and *ab initio* calculations, which lists many of the properties of diatomic molecules. Table 3.2 contains a selection of the binding energy and equilibrium separation of a number of diatomic molecules (Huber and Herzberg, 1979).

The dimer energy and well-depth can easily be fitted to a functional form that would give a curve of roughly the same shape as that shown in Figure 3.1. Two such forms are the so-called 'Lennard-Jones', Lennard-Jones (1924) and 'Morse', Girifalco and Weitzer (1959) potentials. It should be noted that all single-component attractive pair potentials suffer from a drawback in that for modelling crystal structures they are really only applicable to the case of close-packed materials. The global potential minimum is the close-packed configuration. For bcc or diamond lattices, a many-body formulation is required to produce a stable lattice.

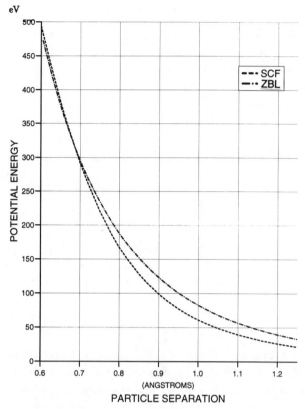

Fig. 3.3 A comparison of the ZBL potential for a two-particle Si–Cu system with self-consistent field (SCF) calculations.

3.3.1 The Lennard-Jones potential

The Lennard-Jones potential was developed to represent noble gases and is based on the premise that atoms in a solid noble gas are only slightly distorted from the stable and closed-shell configuration they possess in the free state. Such small distortions can be described by a dipole interaction and represented by an attractive potential which is proportional to r^{-6}, where r is the interatomic spacing. At short distances the repulsion must be stronger than the attraction and the form of the Lennard-Jones potential is given by

$$V = 4\varepsilon \left[\left(\frac{\sigma}{r} \right)^{12} - \left(\frac{\sigma}{r} \right)^{6} \right]. \tag{3.13}$$

There is no physical reason for the power of 12, other than that it is greater than six to provide repulsion at small particle separation. In this form, $2^{\frac{1}{6}}\sigma$ is the equilibrium spacing and ε the well depth. For Ar, $\varepsilon = 0.01$ eV and $\sigma = 3.4$ Å.

Table 3.2. *Dimer and binding energies and equilibrium separations*

Molecule	Dimer energy (eV)	Equilibrium separation (Å)
Ar_2	−0.01	3.0
Ar_2^+	−1.05	−
As_2	−3.96	2.5
C_2^+	−5.3	1.3
C_2^-	−8.4	1.5
C_2	−6.21	1.3
Cu_2	−2.03	2.3
Ga_2	−1.4	−
$GaAs$	−2.18	−
O_2	−5.35	1.2
Rh_2	−2.9	−
Si_2	−3.21	2.3
SiC	−4.6	1.8
TiN	−4.9	1.6
Xe_2	−0.02	3.6

The potential can also be extended to fit bulk crystal properties, if V is assumed pairwise additive. However, if this is done the dimer properties will not be accurately given. In a crystal the total energy per particle U is given by

$$U = \frac{1}{2} \sum_{R \neq 0} V(R_{ij}), \qquad (3.14a)$$

where R_{ij} is the separation distance between the particles and any other particle in the lattice. The factor $\frac{1}{2}$ is introduced because summing over all atoms in the crystal would give twice the potential energy. It is usual to write R_{ij} as a dimensionless number $\gamma(R_{ij})$ multiplied by the nearest neighbour separation d. Equation (3.14) then becomes

$$U = 2\varepsilon \left[A_{12} \left(\frac{\sigma}{d} \right)^{12} - A_6 \left(\frac{\sigma}{d} \right)^6 \right] \qquad (3.14b)$$

when

$$A_n = \sum_{R \neq 0} \frac{1}{\gamma(R_{ij})^n}.$$

The constants A_{12} and A_6 have been calculated (Jones and Ingham, 1925) for some common lattice types. The values for the simple cubic, bcc and fcc lattices are given in Table 3.3.

Table 3.3. *Values of the Lennard-Jones constants for three different lattice types. Note, however, that the bcc and simple cubic lattices are unstable structures. The unit of distance is taken to be the distance between nearest neighbours.*

	Simple cubic	bcc	fcc
A_6	8.4	12.25	14.45
A_{12}	6.2	9.11	12.13

Table 3.4. *A comparison between experiment and theory for d_0, u_0 and B for fcc Ar*

	Experiment	Theory
d_0 (Å)	3.75	3.71
u_0 (eV)	−0.08	−0.089
B (10^{10} dyne cm^2)	2.7	3.18

To find the nearest neighbour separation r_0 at equilibrium, equation (3.14) is minimised with respect to d. This gives for the fcc lattice

$$d_0 = 1.09\sigma, \quad u_0 = -8.6\varepsilon.$$

Here u_0 is the value of u when $d = d_0$.

The bulk modulus B can also be estimated from the potential for an fcc lattice (Ashcroft and Mermin, 1976) as

$$B = \frac{75\varepsilon}{\sigma}.$$

The pair potential formulation therefore allows a direct comparison with experimental data for the solid noble gases. The agreement is reasonable but the theoretical bulk modulus is 20% larger than that which is observed.

3.3.2 The Morse potential

Another form for the interatomic potential function that satisfies the requirements of Figure 3.1 is the so-called Morse-potential (Girifalco and Weitzer, 1959). Like the Lennard-Jones potential, this potential also contains an

Table 3.5. *Parameters for the Morse potential for Cu*

Potential cut-off distance (nearest neighbour)	D_e	β (Å^{-1})	R_e (Å)
1	0.580	0.01436	2.547
2	0.481	0.01405	2.628
3	0.376	0.01368	2.770
4	0.360	0.01364	2.801
150	0.343	0.01359	2.866

attractive part and a repulsive part. In its original form it had three free parameters, the dimer energy D_e, the equilibrium displacement R_e and a parameter β, which was usually fitted to the bulk modulus of the material. In this section a modified form of the Morse potential is used. The total potential energy E of the solid is assumed to be given by

$$E = \frac{1}{2} \sum_{i \neq j} V_{ij},$$

where

$$V_{ij} = \frac{D_e}{S-1} \left\{ \exp\left[-(2s)^{\frac{1}{2}} \beta \left(r_{ij} - r_e\right)\right] \right. \\ \left. -S \exp\left[-(2/s)^{\frac{1}{2}} \beta \left(r_{ij} - R_e\right)\right] \right\}. \tag{3.15}$$

This reduces to the original Morse potential when $S = 2$, but contains an extra free parameter S, which can be used in the fitting. This form of the potential has also been adapted by a number of authors from the basis of a many-body formulation. In this case the second term in (3.15) is multiplied by a bond-order parameter B, say, which is unity for a two-particle interaction but which depends on the positions of all the neighbours of atom i. This is discussed in detail later in this chapter. The Morse potential has been most frequently used to model the properties of bulk fcc metals. In the form given by equation (3.15), the potential is long-ranged. In MD simulations short-ranged potentials are the most appropriate because large amounts of computing time can be expended calculating the interaction forces. A set of parameters D_e, R_e and β for Cu is contained in Table 3.5, when $S = 2$, for the cases in which the potential cuts off at different distances. These constants were determined by fitting the Morse potential to the lattice spacing, the cohesive energy of Cu and the bulk modulus (Anderman, 1966).

Table 3.6. *A comparison between experimental and calculated data from the Morse potential for Cu*

Potential cut-off distance	c_{11} (10^{12} dyn cm^{-2})	c_{12} (10^{12} dyn cm^{-2})	Vacancy formation energy (eV)	Self-diffusion energy (eV)	Migration energy (eV)
1	2.130	1.065	–	–	–
2	1.856	1.202	–	–	–
3	1.931	1.165	–	–	–
4	1.945	1.158	3.29	4.24	0.95
Experiment	1.762	1.249	1.0	2.05	1.05

Table 3.7. *Morse potential parameters, for some fcc metals*

Morse potential $S = 2$	D_e	R_e (Å)	β (Å^{-1})
Cu	0.481	2.628	0.01405
Rh	0.7595	2.750	0.01566
Mo	0.997	2.800	0.01500
Au	0.560	2.922	0.01637
W	1.335	2.894	0.01200

Some other bulk properties can be calculated and compared with experiments. These are listed in Table 3.6. Although the elastic constants c_{11} and c_{12} are reasonably accurately given, the vacancy formation energy is in error by a factor of three. The dimer binding energy D_e for Cu is 2.03 eV, nowhere near the range of 0.3–0.6 eV given in Table 3.5. In addition, for pairwise additive central forces the identity $c_{12} = c_{44}$ should apply. The observed difference between c_{12} and c_{44} gives some sort of measure of the contribution of the many-body interactions. For Cu, $c_{44} = 0.8177 \times 10^{12}$ dyn cm^{-2}. Even with the extra flexibility of adjusting S, it is not possible to reproduce accurately all the bulk and dimer properties using the pairwise additive Morse potentials. Nonetheless, good simulation results of some measured quantities have been achieved in sputtering studies by using a Morse potential, joined numerically for close particle separation, to a screened Coulomb potential which cuts off after the second nearest neighbour (Harrison, 1988). A list of appropriate parameters is given in Table 3.7 for various metals

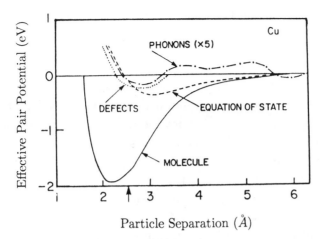

Fig. 3.4 Effective pair potentials for Cu derived from various properties. The nearest neighbour separation in the Cu lattice is shown by the arrow.

The inapplicability of pair potentials to describe all material properties adequately is best shown with reference to Figure 3.4. Here, four different pair binding potentials for Cu are shown, which are fitted to different properties (Carlsson, 1990). Also 'molecule' denotes the Cu_2 binding energy curve and the 'equation of state' is obtained by forcing the potential to fit a quantum mechanically obtained equation of state for the fcc structure. The 'defects curve' is obtained from the vacancy formation energy and other defects properties and the 'phonon' denotes a fit to the phonon spectrum. This illustrates clearly the need to consider many-body potentials for fitting all the important material properties.

3.3.3 Ionic potentials

Both the Morse and the Lennard-Jones potentials are relatively short-ranged, which means that they can readily be incorporated into a computer simulation program that uses neighbour lists. Both the dipole interactions between molecules and the charge interactions between ions are much longer ranged, with the interaction potentials being proportional to r^{-3} and r^{-1} respectively. For a typical simulation of say 2000 particles, the range of the potentials is such that all particles would appear on each others' neighbour list and edge effects would become important. Special techniques are therefore needed to perform the pair interaction sums for ionic materials.

A possible pair potential for ionic materials might therefore take the form

$$V = \text{sgn}\frac{e^2}{r} + \frac{A}{r^m} \qquad A > 0, m > 1 \tag{3.16}$$

per ion pair, where A and m are constants chosen to match the bulk lattice parameters and sgn $= \pm 1$ depending on whether the ions are of similar or opposite charge. The second term involving A and m represents the core repulsion between the particles and, if necessary, V can also be splined to a ZBL or Molière potential for close particle separation. The $4\pi\varepsilon_0$ factor has been omitted from the Coulomb term for simplicity. It is usual to express the interaction energy per ion pair and below it is demonstrated how this energy may be determined for the NaCl lattice.

The cubic NaCl lattice can be represented by two overlapping fcc lattices of anions at sites R and a second fcc lattice of cations displaced through a vector d, which has the magnitude of half the fcc cube side and a direction parallel to the cube edge.

Writing

$$|R| = \gamma(R)d,$$
$$|R + d| = \gamma(R + d)d,$$

where d is the nearest-neighbour spacing, the total Coulombic potential energy U of a single ion can be written

$$U = -\frac{e^2}{d} \left[\frac{1}{\gamma(d)} + \sum_{R \neq 0} \left(\frac{1}{\gamma(R + d)} - \frac{1}{\gamma(R)} \right) \right]. \tag{3.17}$$

This is double the equivalent potential for the Lennard-Jones lattice (equation (3.14)) and a factor of one half should be introduced for consistency. For ionic crystals it is usual to refer to the potential energy per ion pair, rather than per particle. Thus we redefine U given by equation (3.17) to be the potential per ion pair. Unfortunately the series contained in this equation are only conditionally convergent. How to calculate sums such as these will be described later. For the meantime, however, it will be assumed that, if the repulsive core interactions are included, then the total energy per ion pair takes the form

$$U(d) = -\frac{\alpha e^2}{d} + \frac{C}{d^m}.$$

The coefficient α is called the Madelung constant and depends only on the crystal structure. For the NaCl structure $\alpha = 1.7476$.

The equilibrium separation is determined by minimising U. Setting

Table 3.8. *Potential parameters and comparison between experimental* U_{exp} *and theoretical* U_{th} *of energies for Na ionic compounds with the NaCl structure*

Compound	r (Å)	U_{exp} (eV)	U_{th} (eV)	m	C (eVÅ^m)
NaF	2.31	-9.30	-9.36	6.90	510.85
NaCl	2.82	-7.93	-7.80	7.77	3621.31
NaBr	2.99	-7.55	-7.37	8.09	7346.14
NaI	3.24	-7.05	-6.80	8.46	17379.19

$U'(d_0) = 0$ gives

$$C = \frac{\alpha e^2 d_0^{m-1}}{m} = -\frac{\phi_c d_0^m}{m},$$

where $\phi_c = -\alpha e^2/d_0$ is the Coulomb energy and, if d_0 is in ångström units, $\phi_c = -14.4\alpha/d_0$ eV.

Rather than use the free parameter m to fix $U(r)$ to the measured cohesive energy per ion pair it is more usual to determine m in terms of the measured bulk modulus

$$m = 1 + \frac{18Bd_0^4}{\alpha e^2}.$$

This is because the predicted cohesive energy ϕ per ion pair

$$-\frac{\alpha e^2 m}{d_0(m-1)}$$

is only slightly smaller then in magnitude than ϕ_c, which is known to be in reasonable agreement with experiment. For example, for NaCl we find $m = 7.77$, and a difference of 0.13 eV between the experimental and theoretical values of U. A table of values for four Na compounds with the NaCl structure is given in Table 3.8.

More sophisticated ionic potentials have been developed (Fumi and Tosi, 1964, Sangster and Dixon, 1976). A handbook of interatomic potentials for ionic crystals has been compiled by Stoneham (1981). These have generally been pair potentials of a form originally proposed by Huggins and Mayer (1933). Here the interaction potential takes the form

$$V = \text{sgn}\frac{e^2}{r} + B_{ij}\exp{(\alpha_{ij}r)} - \frac{C_{ij}}{r^6} - \frac{D_{ij}}{r^8}. \tag{3.18}$$

The Van der Waals coefficients C_{ij} and D_{ij} have been given by a number

Table 3.9. *Potential parameters for some ionic crystals after Sangster and Dixon (1976). The constant C_{ij} is in units of $eV Å^6$, D_{ij} in units of $eV Å^8$. B_{++} and B_{--} are assumed to be zero*

	α_{+-} ($Å^{-1}$)	B_{+-} (eV)	C_{++}	C_{--}	C_{+-}
NaF	6.9320	593720.3	1.049	10.30	2.809
NaCl	4.4519	25355.8	1.049	72.41	6.991
NaBr	4.0020	14683.5	1.049	122.35	8.739
NaI	3.4658	7422.0	1.049	244.69	11.923

	D_{++}	D_{--}	D_{+-}
NaF	0.50	12.48	2.37
NaCl	0.50	145.44	8.68
NaBr	0.50	280.90	11.86
NaI	0.50	686.64	19.35

of authors. In Table 3.9 the values of Huggins and Mayer (1933) are given. Table 3.9 lists the potential parameters required to use the potential described in (3.18). The potentials are pairwise additive and therefore subject to the restriction that $c_{12} = c_{44}$, thus the elastic properties of the various alkali halides are not always well represented by these potentials.

3.3.4 Lattice sums for ionic potentials

The long-range nature of Coulombic forces means that special techniques are required for their evaluation. The sums contained in equation (3.17) are only conditionally convergent and can therefore be summed to any value whatsoever, depending on the order n to which the summation proceeds. There have been many attempts to perform the lattice sums for periodic structure, the first being due to Ewald (1921).

For a cubic periodic structure the potential at a distance r from a point charge lattice can be written

$$V(r) = \sum_{j=1}^{N} e_j \sum_{i_1=-M_1'}^{M_1} \sum_{i_2=-M_2'}^{M_2} \sum_{i_3=-M_3'}^{M_3} |r - r_j - i_1 L_1 - i_2 L_2 - i_3 L_3|^{-1}, \quad (3.19)$$

where there are N charges per unit cell and charge j at co-ordinate r_j has magnitude e_j. The lattice is assumed neutral so that $\sum_{j=1}^{N} e_j = 0$. The numbers $M'_1, M_1, M'_2, M_2, M'_3$ and M_3 are the number of expected cells in the $\pm L_1, \pm L_2$ and $\pm L_3$ directions. The directions L_1, L_2 and L_3 need not be mutually orthogonal but often are. For a bulk lattice $M'_1 = M_1 = M'_2 = M_2 = M'_3 = M_3 = \infty$. For a semi-infinite lattice, where there is a surface, all the limits are infinite except M_3, which is zero. For a lamina lattice M'_3 and M_3 are finite.

Because the series is conditionally convergent, the value of the potential depends on the shape of the expanding shell of charges about the part of interest, i.e. on the precise way in which the limits tend to infinity. It also depends on the medium surrounding the expanding shell. The results for a sphere surrounded by a good conductor such as a metal and for a sphere surrounded by a vacuum are different (de Leeuw, Perram and Smith, 1980).

For atoms arranged in a regular cubic structure such as the NaCl structure, it is possible to evaluate the lattice sums so that there are no appreciable contributions from charges at the surface. The idea is to divide up the lattice into electrically neutral cubical cells such that the electrostatic interaction energy falls off rapidly with intercellular distance.

Heyes and Van Swol (1981) have shown how the lattice sums may then be summed directly. This technique can be more useful than the Ewald sum technique, especially when surfaces are involved. A lamina potential $V_L(r)$ can be split into two parts:

$$V_L(r) = V_{LD}(r) + V_{LC}(r),$$

where

$$V_{LD}(r) = \sum_{j=1}^{N} e_j \sum_{i_1} \sum_{i_2} \sum_{i_3=-M'_3}^{M_3} |r - r_j - i_1 L_1 - i_2 L_2 - i_3 L_3|^{-1} \qquad (3.20)$$

for $k^2 = i_1^2 + i_2^2 \le i_c^2$ and $V_{LC}(r)$ is the same expression but with $k^2 > i_c^2$. For a square planar lattice $L_1 = L_2 = L$

$$\lim_{i_c \to \infty} V_{LC}(r) = \frac{M_3 + M'_3 + 1}{L^3} \sum_{j=1}^{N} \frac{1}{2} e_j \pi i_c^{-1} \left[(x - x_j)^2 \right.$$

$$\left. + (y - y_j)^2 - 2(z - z_j)^2 \right]. \qquad (3.21)$$

For the NaCl lattice, divided as in Figure 3.5, $V_{LC} = 0$ and the value of the Madelung constant converges rapidly (Table 3.10).

Thus a procedure for implementing a molecular dynamics code for atomic

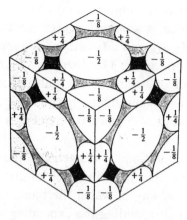

Fig. 3.5 The division of the NaCl structure into cubic cells. In the bulk material an atom at the centre of a face is 'shared' by two cells so its charge is counted as $-\frac{1}{2}$. At a corner an atom is 'shared' by eight cells so its charge is counted as $-\frac{1}{8}$ etc. If the top face is a surface then these numbers become -1 and $-\frac{1}{4}$ respectively.

Table 3.10. *Values of the Madelung constant obtained using the summation technique of Heyes and Van Swol*

k	1	4	4	6
M_3'	1	1	2	2
M_3	1	1	2	2
α	1.503	1.7482	1.7480	1.7476

collision studies in ionic crystals might be to use equations (3.20) and (3.21) to calculate the potential and forces in the undisturbed crystal or to get these from a 'look-up' table. For ions that have been set in motion and have moved from r_j to r_j' a correction term to the potential of the form

$$e_j \left(\frac{1}{|r - r_j'|} - \frac{1}{|r - r_j|} \right)$$

is added.

The book by Allen and Tildersley (1987) describes how the Ewald sum technique can be used to do the full lattice sums for ionic liquids. The book also gives access to the CCP5 program library, from which FORTRAN code for such lattice sums is available.

3.3.5 *Many-body empirical potentials*

The need to model large numbers of material properties has led to the development of many-body potentials. The total potential energy E describing interactions among N identical particles can be resolved into one-body, two-body, ... etc. components as follows:

$$E = \sum_i V_1(r_i) + \sum_{\substack{i,j \\ i<j}} V_2(r_i, r_j) + \sum_{\substack{i,j,k \\ i<j<k}} V_3(r_i, r_j, r_k) + \ldots \qquad (3.22)$$

The three-body and higher order terms contain information about the angles between the bonds. Usually terms greater than the three-body level are not considered because of the computing time required to carry out the calculations. Sometimes the N-body potential is written in a slightly different form as

$$E = \sum_i V_1(r) + \sum_{\substack{i,j \\ i<j}} V_2(r_i, r_j) + \frac{1}{2} \left(\sum_i U(g_2(r_i, r_j), g_3(r_i, r_j, r_k), \ldots) \right). \qquad (3.23)$$

Pseudo-potentials are developed by choosing appropriate mathematical forms for the functions occurring in the above expressions. The mathematical forms are chosen to have a physical meaning but, as with pair potentials, free parameters appear in these forms, which are then used to fit experimental data. It is not the intention here to give a firm theoretical basis for the functional forms to which the experimental data are fitted. Suffice it to state that the precise nature of the mathematical forms chosen is not especially relevant because it has been shown that some forms, chosen by entirely different considerations for semiconductors and metals, are mathematically equivalent (Brenner, 1989). Instead, the parameters for some empirical many-body potentials that have frequently been used in atomic collisions studies will be given and the method by which the fitting can be achieved is outlined.

3.3.6 *Fitting procedures*

Once the functional form has been chosen and the number of free parameters has been set, the question then arises of how the free parameters are chosen and which properties are to be fitted. For atomic collision studies in crystals, a minimum requirement could be the crystal structure, lattice constant, cohesive energy, the bulk modulus and the elastic constants. The vacancy formation energy and the melting temperature are also useful quantities, which are more difficult to fit exactly, but it is possible to test the potential against these once the other parameters have been fitted. For sputtering and deposition studies the dimer and small cluster properties should also be accurately modelled and it is now possible to do this more accurately since many of the energetics and geometrics of small clusters are appearing in the literature as a result of quantum mechanical SCF calculations. In surface studies the potentials should also fit the known surface reconstructions. The elastic constants should also be reproduced and the formulae for these in terms of the interatomic potential are given by Johnson (1972) and Ackland (1992). However, it is often just as easy to calculate these numerically by small distortions of the crystal, calculating the change in energy and performing a numerical differentiation.

Choosing a parameter set to model all the properties accurately is not an especially easy task. One 'brute force' way in which this can be achieved is first to fit exactly as many properties of the potential as possible. Then, if the form of the potential has sufficient free parameters, an objective function C could be defined, which has to be minimised:

$$C = \sum_{k=1}^{M} \alpha_k \left(pr^k_{\mathrm{KNOWN}} - pr^k_{\mathrm{CALC}} \right)^2 + \mathrm{PF},$$

where pr^k_{KNOWN} is the known kth property to be fitted, pr^j_{CALC} is the property given by the calculation and α_k are positive weights. The term PF is a positive penalty function. An example in which the penalty function might be needed could be in fitting the required crystal lattice structure, where PF is defined to be positive if another lattice structure would give a lower potential energy. The choice of the weights α_k and the penalty function PF depends on the problem in hand and can really only be done by trial and error. The choice of optimisation algorithm is also fairly crucial. Downhill local optimisation algorithms are not appropriate since they nearly always approach a local minimum and ideally we wish to find the global minimum where $C = 0$. In our experience a variant of the controlled random search algorithm of Price (1983) has proved attractive. It improves on a pure Monte Carlo random search and is sufficiently fast for the small parameter sets usually required. However, fitting procedures do not necessarily involve global optimisation, see for example Andersson, Johnston and Murrell (1994).

3.3.7 The Finnis–Sinclair method and the embedded atom method for metals

In the embedded atom method (EAM), the energy of each atom is computed from the energy needed to embed the atom in the local electron density provided by the other atoms in the metal. In the original EAM method of Daw and Baskes (1983, 1984), the electron density is approximated by the superposition of atomic electron densities and so computationally the EAM energy can be evaluated with about the same amount of work as pair potentials.

The EAM energy of an atom labelled i is written as

$$U_i = -f(\rho_i) + \frac{1}{2}\sum_j V_{ij}, \qquad (3.24)$$

where V is a central pair potential, f an embedding function and the electron density ρ_i is approximated by the superposition of atomic densities

$$\rho_i = \sum_j \phi_{ij} \qquad (3.25)$$

and ϕ_{ij} is a function only of the interatomic distance r_{ij} between atoms i and j. Daw and Baskes use a fitting procedure for f, as a function of ρ and ϕ_{ij}. The function ϕ_{ij} is written as

$$\phi_{ij} = Z_i Z_j / r_{ij}; \quad Z(r) = Z_0(1 + \beta r^\nu)e^{-\alpha r},$$

where α, β and ν are adjustable parameters and Z_0 is the number of valence electrons. The parameters α, β and ν are given for various fcc metals but the function F and the atomic density ρ are not parameterised and are given numerically. To incorporate this potential into a MD simulation requires input tables for ρ and F. Other authors (Ackland *et al.* 1987, Voter and Chen, 1987) have parameterised fcc potentials in a convenient form to use, which fits the bulk properties of fcc metals equally well. Ackland has used the Finnis–Sinclair (Finnis and Sinclair, 1984) tight-binding form of f i.e. $F(\rho) = -A\rho^{\frac{1}{2}}$ together with parameterised fits to ϕ_{ij} to give potentials valid for fcc, bcc and hcp metals that are easily programmed.

For the fcc and hcp metals the potential takes the functional forms

$$V(r) = \sum_{k=1}^{6} a_k(r_k - r)^3 \Theta(r_k - r); \quad r_1 > r_2 \ldots > r_6, \qquad (3.26)$$

$$\phi(r) = \sum_{k=1}^{2} A_k(R_k - r)^3 \Theta(R_k - r); \quad R_1 > R_2, \qquad (3.27)$$

where $\Theta(x)$ is the Heaviside step function

$$\Theta(x) = \begin{cases} 0 & x < 0 \\ 1 & x > 0. \end{cases}$$

The quantities r_1 and R_1 represent the cut-off radii for V and ϕ and the relevant parameters are given in Table 3.11.

Table 3.11. *Fitted coefficients. The coefficients for V and φ are in electronvolts and values of r_k and R_k are in units of the lattice parameter. For nickel the stacking fault energy is not fitted.*

Coefficent	Cu	Ag	Au	Ni	Ti (hcp)
a_1 (eV)	29.059214	20.368404	29.059066	29.057085	−57.099097
a_2 (eV)	−140.05681	−102.36075	−153.14779	−76.04625	80.735598
a_3 (eV)	130.07331	94.31277	148.17881	48.08920	−21.761468
a_4 (eV)	−17.48135	−6.220051	−22.20508	−25.96604	−10.396479
a_5 (eV)	31.82546	31.08088	72.71465	79.15121	74.515028
a_6 (eV)	71.58749	175.56047	199.26269	0	35.921024
R_1 (LU)	1.2247449	1.2247449	1.1180065	1.2247449	1.22
R_2 (LU)	1.0000000	1.0000000	0.8660254	1.1180065	1.05
A_1 (eV)	9.806694	1.458761	21.930125	60.537985	39.795927
A_2 (eV)	16.774638	42.946555	284.99631	−80.102414	−40.4061305
r_1 (LU)	1.2247449	1.2247449	1.2247449	1.2247449	1.22
r_2 (LU)	1.1547054	1.1547054	1.1547054	1.1547054	1.20
r_3 (LU)	1.1180065	1.1180065	1.1180065	1.1180065	1.12
r_4 (LU)	1.0000000	1.0000000	1.0000000	1.0000000	0.93
r_5 (LU)	0.8660254	0.8660254	0.8660254	0.8660254	0.80
r_6 (LU)	0.7071068	0.7071068	0.7071068	0.7071068	0.707107

It should be noted that the dimer energies are not fitted exactly with the fcc and hcp potentials given above. These potentials have been tested to ensure that the fcc and hcp structures are the minimum energy configurations. In addition, satisfactory agreement was reported when testing the potentials with experimental data that were not fitted, such as the monovacancy formation energy.

Ackland and Thetford's modification to the Finnis–Sinclair bcc potential has the form (Ackland and Thetford, 1987)

$$U_i = A\rho_i^{\frac{1}{2}} + \frac{1}{2}\sum_j V_{ij},$$

$$\rho_i = \sum_j \phi_{ij},$$

$$\phi = \Theta(d - r)(r - d)^2,$$

$$V = \Theta(c - r)(r - c)^2(c_0 + c_1 r + c_2 r^2) + \Theta(b_0 - r)B(b_0 - r)^3 e^{-\alpha r}.$$

Table 3.12. *Parameters for potentials for some bcc metals*

	A (eV)	B (eV Å^{-3})	α (Å^{-1})	b_0 (Å)	c (Å)	c_0
V	2.010637	23.0	0.5	2.6320	3.80	−0.8816318
Nb	3.013789	48.0	0.8	2.8585	4.20	−1.5640104
Ta	2.591061	91.0	1.05	2.8629	4.20	1.2157373
Mo	1.887117	1223.0	3.9	2.7255	3.25	43.4475218
W	1.896373	90.3	1.2	2.7411	3.25	47.1346499

	c_1	c_2	d (Å)
V	1.4907756	−0.3976370	3.692767
Nb	2.0055779	−0.4663764	3.915354
Ta	0.0271471	−0.1217350	4.076980
Mo	−31.9332978	6.0004249	4.114825
W	−33.7665655	6.2541999	4.400224

The parameters for five bcc metals are given in Table 3.12. The potential is fitted to the experimental values of the lattice constant and cohesive energy, the elastic constants c_{11}, c_{12} and c_{44} and the Cauchy pressure.

The embedded atom method has also been applied to Si, which has highly directional covalent bonding. Here a simple analytic form for $f(\bar{\rho})$ was used:

$$f(\bar{\rho}) = \frac{\bar{\rho}}{\bar{\rho}_0} \ln \left(\frac{\bar{\rho}}{\bar{\rho}_0} \right),$$

where $\bar{\rho}_0$ is the equilibrium bulk density at an atomic site. This form for f is motivated by the experimental observation of a universal correlation between the number of bonds and the bond length. The density $\bar{\rho}$ was taken to be angularly dependent in the form

$$\bar{\rho} = \sum_{j \neq i} \rho(r_{ij}) - c_0 \sum_{j,k \neq i} (1 - 3 \cos^2 \theta_{jik}) \rho(r_{ij}) \rho(r_{ik}), \qquad (3.28)$$

where θ_{jik} is the bond angle subtended at atom i by atoms j and k and c_0 is a fitting parameter. The second term in the above expression is a three-body term, which is zero in the diamond lattice structure. Again the EAM potential for Si depends on the function $\rho(r_{ij})$, which is numerically given. Thus the Si potential is less convenient to use than some other forms and so we briefly mention two of the most popular which have been used in simulation studies.

3.3.8 Covalent materials

Because of the importance of silicon technology, a great deal of effort has been expended in determining empirical potentials for use in modelling Si. We report on two of the most popular here but many more have appeared in the literature.

The Stillinger–Weber potential. The Stillinger–Weber (Stillinger and Weber, 1985) potential is an approximation to equation (3.20) with $V_1 = 0$

$$v_2 = \varepsilon f_2(\bar{r}_{ij}); \quad V_3 = \varepsilon \left[f_3(\bar{r}_{ij}, \bar{r}_{ik}, \theta_{jik}) + f_3(\bar{r}_{ji}, \bar{r}_{jk}, \theta_{ijk}) \right.$$

$$\left. + f_3(\bar{r}_{ki}, \bar{r}_{kj}, \theta_{ikj}) \right], \tag{3.29}$$

where $\bar{r}_{ij} = r_{ij}/\sigma$ and f_2 and f_3 are given by

$$f_2(r) = \begin{cases} A(Br^{-p} - r^{-q}) \exp\left[(r-a)^{-1}\right] & r < a \\ 0 & r > a, \end{cases} \tag{3.30}$$

$$f_3(r, s, \theta) = \begin{cases} \lambda \exp\left[\gamma(r-a)^{-1} + \gamma(s-a)^{-1}\right] \left(\cos\theta + \tfrac{1}{3}\right)^2; & r, s < a \\ 0; & r, s \geq a. \end{cases} \tag{3.31}$$

The parameter set is given by

$$A = 7.049556277, \qquad\qquad B = -.6022245584,$$
$$p = 4, \qquad q = 0, \qquad\qquad a = 1.8,$$
$$\lambda = 21.0, \quad \gamma = 1.2, \qquad\qquad \sigma = 2.0951 \text{ Å}, \quad \varepsilon = 5.56262 \text{ eV}.$$

In the diamond lattice structure the three-body term V_3 is identically zero. If the bond angle θ_{jik} is not the ideal tetrahedral angle of $109°28'$ then this term is positive, thus increasing the potential energy. The effect therefore of this term is to stabilise the diamond lattice as the minimum energy configuration. This potential cuts off at 3.77 Å so, although the three-body terms require more computing time to evaluate, its relatively short range makes it very convenient for inclusion in MD simulations. The potential fits the cohesive energy and lattice constant for Si, the melting point and liquid structure. It fits neither dimer properties nor the elastic properties of Si.

Stillinger and Weber have also developed an F potential, which can be used in conjunction with the Si potential to model F etching of Si (Stillinger and Weber, 1989).

Bond order potentials. An alternative functional fitting form for Si is based on the premise that atomic coordination is the main variable determining the bonding properties of silicon in different structures. In this formulation the total energy of the system is written as a sum over atomic sites in the form

$$\phi = \frac{1}{2} \sum_{i \neq j} f_{\rm c}(r_{ij})(V_{\rm R}(r_{ij}) - B_{ij}V_{\rm A}(r_{ij})), \tag{3.32}$$

where $V_{\rm R}$ is a repulsive term, $V_{\rm A}$ is an attractive term, $f_{\rm c}$ is a switching function, which cuts off the potential smoothly at suitably large values of r_{ij}, and B_{ij} is a many-body term related to bond order that depends on the positions of atoms i and j and the neighbours of atom i. In the Tersoff formulation (Tersoff, 1988) the function B_{ij} is given by

$$\begin{aligned} B_{ij} &= (1 + \alpha^n \xi_{ij}^n)^{\frac{1}{2n}}, \\ \xi_{ij} &= \sum_{k \neq i,j} f_{\rm c}(r_{ik})g(\theta_{ijk})\exp[\lambda_3^3(r_{ij} - r_{ik})^3], \\ g(\theta) &= 1 + \frac{c^2}{d^2} - \frac{c^2}{[d^2 + (h - \cos\theta)^2]}. \end{aligned} \tag{3.33}$$

The function $f_{\rm c}$ is given by

$$f_{\rm c}(r) = \begin{cases} 1 & r \leq R - D \\ \frac{1}{2} - \frac{1}{2}\sin[\pi(r - R)/(2D)] & R - D \leq r \leq R + D \\ 0 & r \geq R + D. \end{cases}$$

The functions $V_{\rm R}$ and $V_{\rm A}$ take the forms

$$V_{\rm R}(r) = A\exp(-\lambda_1 r); \qquad V_{\rm A}(r) = B\exp(-\lambda_2 r).$$

Two parameter sets have been given by Tersoff, which fit the lattice spacing for Si, the cohesive energy and the diamond lattice structure as the minimum energy configuration. The appropriate parameter sets are given in Table 3.13. Note that values of R, D, α and λ_3 are not systematically optimised in this fit.

The Tersoff formulation has been used to give parameter sets for C, SiC (Tersoff, 1988, 1989) and GaAs (Smith, 1992). Brenner (1990, 1992) has also used the basic Tersoff formulation to fit potentials to hydrocarbons, primarily for use in modelling the growth of diamond-like films.

3.4 The overlap potential

The applicability of the attractive potentials described in Section 3.3 to model cohesion and bonding and the repulsive potentials in Section 3.2 to model

Table 3.13. *The parameter sets of two Si potentials given by Tersoff.*
Si(1) has good surface properties but underestimates c_{44} by an order
of magnitude. Si(2) fits the elastic constants but Si surface properties
are less accurately described

	Si(1)	Si(2)
A (eV)	47×10^3	1.8308×10^3
B (eV)	9.5373×10^1	4.7118×10^2
λ_1 (Å^{-1})	3.2394	2.4799
λ_2 (Å^{-1})	1.3258	1.7322
α	0.0	0.0
β	$3.3675 \times 10{-1}$	1.0999×10^{-6}
n	2.2956×10^1	7.8734×10^{-1}
c	4.8381	1.0039×10^5
d	2.0417	1.6218×10^1
h	0.0000	-5.9826×10^{-1}
λ_3 (Å^{-1})	1.3258	1.7322
R (Å)	3.0	2.85
D (Å)	0.2	0.15

scattering suggests that the two types of potential should be numerically
joined in order to be applicable for use in cascade studies over a wide energy
range. Harrison (1988) used a cubic spline to join the Morse potential to
the repulsive screened Coulomb potential. So the resulting potential would
be described by

$$\phi = \begin{cases} \text{Screened Coulomb} & 0 < r < r_a \\ \text{Cubic spline} & r_a \leq r < r_b \\ \text{Morse potential} & r_b \leq r < r_c \\ 0 & r > r_c. \end{cases}$$

There is some flexibility over the choice of r_a and r_b but these are generally
fixed at the start. Harrison also used the β parameter appearing in the
Morse potential as another variable and the screening length in the Molière
potential. This usually gives a smooth join between the potentials but in some
cases the cubic spline can produce oscillations in the potential curve over
the joining region, which appear unphysical and aesthetically displeasing.
Instead, another fitting procedure has been adopted, which works well for
the case of the Tersoff Si potential and which will generalise to fit other
many-body or pair binding potentials (Smith, Garrison and Harrison, 1989).

The procedure is to join the forces from the repulsive part of the Tersoff potential $V_R(r_{ij})$ to an inner screened Coulomb potential with a function of the form

$$F = \exp(ar_{ij} + b)$$

where a and b are fitting constants. The distances r_a and r_b are also regarded as free parameters. The fitting procedure is as follows. First choose r_a and r_b to be reasonable values. The function $\exp(ar_{ij} + b)$ is then fitted to the forces of the screened Coulomb potential at r_a and the forces at r_b. The constant of integration fits V_R at $r = b$. This will not in general give the correct value of the potential at r_a and so r_a and r_b can be adjusted until the correct value of $V(r_a)$ is given. For example, for the Tersoff Si(2) potential defined in Table 3.12, splined to a Molière potential with a screening length of 0.8 times the Firsov value, the values of $r_a = 0.353$ Å and $r_b = 0.543$ Å give a smooth fit.

Collision cascades can often be sensitively dependent on the overlap regime. Unfortunately this region is often not modelled well. This is an area where further research is required.

3.5 Conclusion

The semi-empirical potentials described in this chapter are not a complete list in any sense. They have been included because of their relative ease of implementation both in MD and in BC codes and also because of the range of materials and structures that are covered. For a thorough discussion of binding potentials see the review article by Carlsson (1990).

Binary collision computer codes for modelling atomic collisions rarely include the binding part of the potential. Indeed, for studying phenomena such as ion implantation there is no need to do this since the mean ranges and cascade spreading are governed primarily by the energy loss (both nuclear and electronic) in the energetic part of the cascade. For such calculations the ZBL potential is generally regarded as the best 'mean' potential to use to model the atomic collisions. This is the potential used in most versions of TRIM (Biersack and Eckstein, 1984) and in the CRYSTAL code (Chakarov and Webb, 1994).

The Molière potential is also commonly used, sometimes with an adjustable screening length, depending on the particular projectile–target compositions. However, if collision cascades are to be followed to their full conclusion, then the recrystallisation which takes place at the end of the cascade can only be modelled well if the cohesion in the crystal is accurately described. This means using many-body empirical potentials such as those described above to describe the crystal bonding. Although such potentials can never provide the quantitative accuracy of fully quantum mechanical calculations, increasingly sophisticated descriptions of this kind will almost certainly reveal new insights into material properties.

Of course, a proper dynamical treatment of cascades would be an MD simulation based on forces derived by *ab initio* calculations. Combined calculations of this kind achieved notable success when Car and Parrinello (1985) simulated an eight-atom Si crystal. Since then, with the use of parallel computing, systems containing hundreds of atoms have been studied. For example, Gillan (1993) has reported that a system of 400 atoms has simulated the 7×7 reconstruction of the Si$\{111\}$ surface. The bond order potential method, Sutton (1992), also offers intriguing possibilities. However, the realistic study of atomic collisions at higher energies is still beyond the scope of even the most massively parallel computer.

There is no doubt, however, that combined MD/*ab initio* calculations will increase in popularity and reveal new insights into atomic scale dynamical phenomena over the next few years.

References

Ackland, G. J. (1992). *Phil. Mag.* A **66** 917.
Ackland, G. J. Tichy, G. Vitek, V. and Finnis, M. W. (1987). *Phil. Mag.* A **56** 735.
Ackland, G. J. and Thetford, R. (1987). *Phil. Mag.* A **56** 15.
Allen, M. P. and Tildersley, D. J. (1987). *Computer Simulation of Liquids*, Clarendon Press, Oxford.
Anderman, A. (1966). Report AFCRL-66-688, Atomics International, Canoga Park.
Andersson, K. M., Johnston, R. L. and Murrell, J. M. (1994). *Phys. Rev.* **B 49** 3089.
Ashcroft, N. W. and Mermin, N. D. (1976). *Solid State Physics*, Holt, Rinehart and Winston, Philadelphia.
Biersack, J. P. and Eckstein, W. D. (1984). *Appl. Phys.* A **34** 37.
Biersack, J. P. and Haggmark, L. G. (1980). *Nucl. Instrum. Meth.* B **174** 257.
Bohr, N. (1948). *Kgl. Dansk. Vid. Selsk. Mat. Fys. Medd.* **18**.
Brenner, D. W. (1989). *Phys. Rev. Lett.* **63** 1022.
Brenner, D. W. (1990). *Phys. Rev.* B **42** 9458.
Brenner, D. W. (1992). *Phys. Rev.* B **46** 1948.
Car, R. and Parrinello, M. (1985). *Phys. Rev. Lett.* **56** 2471.
Carlsson, A. E. (1990). *Solid State Phys.* **43** 1.
Chakarov, I. and Webb, R. P. (1994). *Rad. Eff. Defects Solids* **130/131**, 447.
Daw, M. S. and Baskes, M. I. (1983). *Phys. Rev. Lett.* **50** 1285.
Daw, M. S. and Baskes, M. I. (1984). *Phys. Rev. Lett.* B **29** 6443.
de Leeuw, S. W., Perram, J. W. and Smith, E. R. (1980). *Proc. Roy. Soc* A **373** 27.
Ewald, P. (1921). *Ann. Phys.* **64** 253.
Finnis, M. W. and Sinclair, J. E. (1984). *Phil. Mag.* A **50** 45.
Firsov, O. B. (1957). *Zh. Eksperim. Teor. Fyz.* **33** 696.

Fumi, G. G. and Tosi, M. P. (1964). *J. Phys. Chem. Solids* **25** 31.

Gillan, M. (1993). *New Scientist* 3 April 35.

Girifalco, L. A. and Weitzer, V. G. (1959). *Phys. Rev.* **114** 687.

Harrison, D. E. (1988). *CRC Crit. Rev. Solid St. Mater. Sci.* **14** 51.

Hehre, W., Ditchfield, R., Radom, L. and Pople, J. A. (1970). *J. Am. Chem. Soc.* **92** 4796.

Heyes, D. M. and Van Swol, F. (1981). *J. Chem. Phys.* **75** 5051.

Huber, K. P. and Herzberg, G. (1979). *Constants of Diatomic Molecules*, Van Nostrand Reinhold, New York.

Huggins, M. L. and Mayer, J. E. (1933). *J. Chem. Phys.* **1** 643.

Johnson, R. A. (1972). *Phys. Rev.* B **6** 2094.

Jones, J. E. and Ingham, A. E. (1925). *Proc. Roy. Soc.* A **107** 636.

Lennard-Jones, J. E. (1924). *Proc. Roy. Soc.* A **106** 441.

Price, W. L. (1983). *J. Opt. Theory Appl.* **40** 333.

Sangster, M. J. L. and Dixon, M. (1976). *Adv. Phys.* **25** 247.

Smith, R. (1992). *Nucl. Instrum Meth.* B **67** 335.

Smith, R., Harrison, D. E. and Garrison, B. J. (1989). *Phys. Rev.* B **40** 93.

Stillinger, F. H. and Weber, T. A. (1985). *Phys. Rev.* B **31** 5262.

Stillinger, F. H. and Weber, T. A. (1989). *Phys. Rev. Lett.* **62** 2144.

Stoneham, A. M. (1981). *Handbook of Interatomic Potentials 1, Ionic Crystals*, AERE-R-1598, AERE, Harwell, Oxon.

Sutton, A. P. (1992). *Electronic Structure of Materials*, Oxford Science Publications, Oxford.

Tersoff, J. (1988). *Phys. Rev. Lett.* **61** 2879.

Tersoff, J. (1989). *Phys. Rev.* B **39** 5566.

Torrens, I. M. (1972). *Interatomic Potentials*, Academic Press, New York.

Voter, A. D. and Chen, S. P. (1987). *Mater. Res. Symp. Proc.* **82** 175.

Ziegler, J. F., Biersack, J. P. and Littmark, U. (1985). *The Stopping and Range of Ions in Solids*, Pergamon, New York.

4

Electronic energy loss models

4.1 Introduction

During their passage through matter ions interacts not only with the atoms from the lattice but also with the electrons. For collisions with electrons, in most cases, energy is transferred from the moving particle to the electrons in the lattice. Processes involving energy transfer to the electrons are called electronic energy losses. These losses could involve elastic scattering of electrons by the moving particle and the collisions could also involve inelastic losses.

In an elastic collision, the total kinetic energy of the system of colliding particles is the same before and after the collision and there is no change in the internal excitation energies, neither of the atom nor of the electrons. The process may be summarized as follows:

$$e_{E_i} + e_A = \text{constant},$$

where e_{E_i} represents the electron with initial energy E_i and e_A is the atom energy.

If an electron is scattered through an angle θ there is a momentum transfer from the electron to the incident particle. Figure 4.1 shows the geometry of the colliding particles in the centre-of-mass (CM) co-ordinate system, where the ion is almost at rest (because the ion mass is so much greater than that of the electron). The momentum of the electron before the collision is mv. After scattering, the electron experiences a change of its forward momentum by the amount

$$\Delta P = mv \left(1 - \cos \theta\right).$$

When $\theta > 90°$, backward scattering occurs.

In an inelastic collision, it is usually assumed that the total momentum is conserved but the total kinetic energy decreases as a result of the excitation of

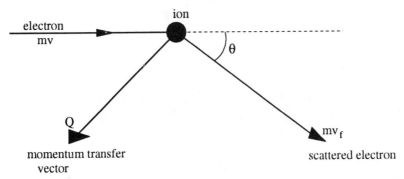

Fig. 4.1 A schematic diagram of an electron–atom collision event in a CM co-ordinate system. An electron with initial momentum mv collides with an almost stationary atom, resulting in scattering through angle θ and final momentum mv_f.

one or both of the collision particles. Ions can be slowed down by collisions with the electrons of other atoms or electrons of incident kinetic energy E_i (in the CM co-ordinate system) may excite an atom from an initially bound state to another bound state of higher electronic energy. This can occur only when $E_i > \delta E$, where δE is the energy separation of the two bound states

$$e_{E_i} + e_A = e_{A^*} + e_{(E_i - \delta E)} \; .$$

How these processes lead to energy loss will be discussed later in this chapter.

Thus, the stopping of a moving particle in the solid is usually considered to consist of two separate processes – recoils with the atomic nuclei and electronic processes. The electronic energy losses are also called electronic stopping. In Figure 4.2 the dependence of the electronic stopping $(-dE/dx)$ is plotted as a function of the velocity of an ion with atomic number Z_1 moving in a solid with atomic number Z_2. In region III, corresponding to a high energy of the moving particle, the electron excitation energy in the solid has some average value $\langle \Delta E \rangle$, which does not depend on the ion velocity v, and the electronic stopping can be represented as a probability for inelastic scattering multiplied by $\langle \Delta E \rangle$. Here the moving particle is fully ionised. Detailed calculations of energy loss in this high-energy region have been performed by Bethe (1930) and Bloch (1933). The first term in the approximation gives the stopping proportional to v^{-2}.

As the energy decreases the effect of the polarisation charge of the particle becomes important. In region II the stopping power displays a maximum. One important aspect of the intermediate energy stopping power is its dependence on charge-state effects. It is primarily in this region that the greatest range of ionization states will exist. In region I, corresponding to very low energies, we have $v < v_B Z_1^{2/3}$ (where $v_B = e^2/\hbar = 2.1877 \times 10^{16}$ Å

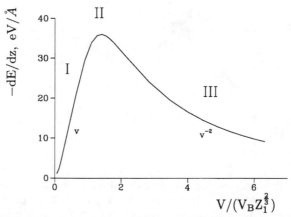

Fig. 4.2 The dependence of the electronic stopping $-\mathrm{d}E/\mathrm{d}x$ on the velocity of He ions in silicon.

s^{-1} is the Bohr velocity)[1] and the electronic stopping can be treated as a retarding force proportional to v.

4.2 Electronic stopping for low energies

4.2.1 Firsov's semi-classical model (local)

In Firsov's model (Firsov, 1959), the estimation of the electronic energy loss per collision is based on the assumption of a quasi-classical picture of the electrons i.e. an average energy of excitation of electron shells, and electron distribution and motion according to the Thomas–Fermi model of the atom (see Chapter 3). In this quasi-classical picture, the transfer of energy, ΔE, from the ion to the atom is due to the passage of electrons from one particle to the other. This results in a change of the momentum of the ion, which arises from the retarding force acting on the ion. When the ion moves away from the atom, the electrons return. However, there is no back transfer of momentum because the electrons fall into higher energy levels.

According to the Thomas–Fermi model, the mean velocity of the electrons in the atom is

$$u(r) = \frac{3}{4m}(3\pi^2)^{1/3}\hbar n^{1/3}(r), \tag{4.1}$$

where $n(r)$ is the electron density at a distance r from the nucleus and m is the electron mass. Let us introduce a surface S dividing the regions of

---

[1] As long as $v \ll v_{\mathrm{F}}$, where $v_{\mathrm{F}} = 1.919v_{\mathrm{B}}/[3/(4\pi na_{\mathrm{B}}^3)]^{1/3}$ is the Fermi velocity of the conduction electrons in the solid, the ions move too slowly to excite the target atom cores and the stopping power is determined predominantly by excitation of conduction electrons. In the above, n is the electron density and a_{B} the Bohr radius.

action of the potentials of the ion and the atom, Figure 4.3. The transverse component of the electron velocity averaged over all directions in space is

$$\langle u_\perp \rangle = \frac{\int_0^{2\pi} \int_0^{\pi/2} u \cos\theta \sin\theta \, d\theta \, d\varphi + \int_0^{2\pi} \int_{\pi/2}^{\pi} u \cos\theta \sin\theta \, d\theta \, d\varphi}{\int_0^{2\pi} \int_0^{\pi} \sin\theta \, d\theta \, d\varphi}$$

$$= \frac{(u \cdot 2\pi/2) - (u \cdot 2\pi/2)}{2 \cdot 2\pi} = 0. \tag{4.2}$$

In the last expression the first term corresponds to transfer of the momentum of the electron from the atom to the ion. Passing through S, the electrons strongly interact with the field of the corresponding atom, losing their initial momentum and, on the average, assuming a momentum corresponding to the velocity of the atom. The electron flux density in one direction, through the element of area dS at distance r from the centre of the atom is $n(r)[u(r)/4] \, dS$, so the total momentum transfer at any moment, i.e. the force acting on the ion, is

$$d\boldsymbol{F} = m\dot{\boldsymbol{R}} n(r)[u(r)/4] \, dS, \tag{4.3}$$

or

$$\boldsymbol{F} = m\dot{\boldsymbol{R}} \int_S n(r)[u(r)/4] \, dS, \tag{4.4}$$

where $\dot{\boldsymbol{R}}$ is the relative velocity of the two interacting atoms and u is the mean value of the electron velocity defined in 4.1. The total work for slowing down the ion (in other words the electron excitation energy) is

$$\Delta E = m \int \left(\int_S \frac{n(r)u(r)}{4} \, dS \right) \dot{\boldsymbol{R}} \, d\boldsymbol{R}. \tag{4.5}$$

Several assumptions are made to evaluate further the integral in equation (4.5). First, $n(r)$ and $u(r)$ are related to the potential $V_0(r)$ in the intervening space, S, in the same way as they are in the Thomas–Fermi atomic model (see equations (3.6) and (4.1)), i.e., $n(r) \propto V_0(r)^{3/2}$ and $u(r) \propto V_0(r)^{1/2}$:

$$n(r) = \frac{1}{3\pi^2 \hbar^3} (2me V_0)^{3/2}, \tag{4.6}$$

$$u(r) = \frac{3}{4m} (2me V_0)^{1/2}, \tag{4.7}$$

where e is the electronic charge.

In addition, it is assumed that $|\dot{\boldsymbol{R}}| \ll u$. If this were not the case, the momentum transfer would be much smaller. If we limit the discussion only to small-angle scattering, in which the motion of the nuclei is assumed to

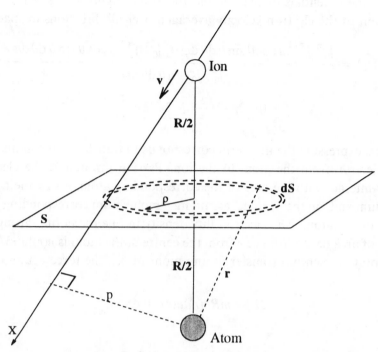

Fig. 4.3 The surface S, dividing the regions of action of the potentials of the ion and the atom.

be rectilinear and uniform, then $\dot{R}\,dR = (\dot{R} \cdot d\mathbf{R}_{\parallel}) = v\,dx$, where $v = |\dot{R}|$, $dx = \dot{R} \cdot d\mathbf{R}_{\parallel}/|\dot{R}|$ and $d\mathbf{R}_{\parallel}$ is a vector parallel to the velocity $\mathbf{v}$.

Substituting (4.6) into (4.5) and taking into account the above assumptions, the electron excitation energy for the collision becomes

$$\Delta E = \frac{3}{16}(3\pi^2)^{1/3}\hbar v \int_{-\infty}^{\infty} \left(\int_S [n(r)]^{4/3} dS \right) dx, \qquad (4.8)$$

where S is the entire plane. Thus, the problem is reduced to specifying the potential in the space between the atoms. Firsov's further assumption is that $V_0(r)$ is given by

$$V_0(r) = \frac{Z_1 Z_2}{r} \chi \left[1.13(Z_1 + Z_2)^{1/3} r \right], \qquad (4.9)$$

where Z_1 and Z_2 are the atomic numbers of the ion and the atom, χ is the Thomas–Fermi screening function (Marich, 1950), and r is the distance in a.u. ($a_B = 1$ a.u. $= 0.529$ Å) between the atom and the surface S. Equation 4.9 is equivalent to equations 3.11, except that atomic rather than S.I. units are used here.

From Figure 4.3 we have the relations

$$R^2 = p^2 + x^2,$$

$$r^2 = \frac{R^2}{4} + \rho^2 = \frac{p^2}{4} + \frac{x^2}{4} + \rho^2.$$

Finally, for the electronic energy loss we obtain

$$-\Delta E = \frac{3}{16}(3\pi^2)^{1/3}\hbar v \int_{-\infty}^{\infty} \int_0^{\infty} n \left[\left(\frac{p^2}{4} + \frac{x^2}{4} + \rho^2 \right)^{\frac{1}{2}} \right]^{\frac{4}{3}} 2\pi\rho \, d\rho \, dx, \quad (4.10)$$

where the negative sign is included in front of ΔE to signify energy loss.

Firsov performed the integration using the expression for the electron density from the Thomas–Fermi model and equation (4.9):

$$-\Delta E = \frac{me^4}{2\hbar^2} \frac{0.7(Z_1 + Z_2)^{5/3}}{[1 + 0.16(Z_1 + Z_2)^{1/3}(p/a_B)]^5} \frac{v}{v_B}, \quad (4.11)$$

or, with evaluation of the physical constants,

$$-\Delta E = \frac{4.3 \times 10^{-8}(Z_1 + Z_2)^{5/3}v}{[1 + 0.31(Z_1 + Z_2)^{1/3}p]^5} \text{eV}, \quad (4.12)$$

where v is the relative velocity of the atoms in cm s^{-1}.

Expressing the velocity of the moving particle in electronvolts, since $v \text{ (cm s}^{-1}) = 1.389 \times 10^6 \left[E \text{ (eV)}/m_1 \right]^{\frac{1}{2}}$, we obtain

$$-\Delta E = \frac{0.05973(Z_1 + Z_2)^{5/3}(E/m_1)^{\frac{1}{2}}}{\left[1 + 0.31(Z_1 + Z_2)^{1/3}R_0(p, E)\right]^5} \text{eV}. \quad (4.13)$$

Here, R_0 is the distance of closest approach in ångström units. This is approximately equal to the impact parameter, p, in the case of small-angle collisions. The replacement of p by R_0 is a correction introduced by Robinson and Torrens (1974). E is the energy of the moving atom (the ion) in electronvolts and m_1 is its mass in a.m.u. In a binary collision, the scattering angles are affected by the inelastic energy loss ΔE (equation (2.18)).

To calculate the electronic stopping cross-section for random media, one must integrate the energy transferred in one interaction between two atoms over all possible impact parameters (more precisely, the distance of closest approach, R_0)

$$S_e(E) = 2\pi \int_0^{\infty} \Delta E \, R_0 \, dR_0. \quad (4.14)$$

Performing this integration using Firsov's excitation energy ΔE we obtain

$$S_e(v) = 0.234 \times 10^{-6}(Z_1 + Z_2)v \text{ eV Å}^2 \qquad (4.15)$$

or

$$S_e(E) = 0.325(Z_1 + Z_2)\left(\frac{E}{m_1}\right)^{\frac{1}{2}} \text{ eV Å}^2. \qquad (4.16)$$

Example 4.1. Calculate the electronic stopping cross-sections as a function of atomic number of the incident particle moving in the middle of a $\langle 110 \rangle$ channel in silicon. The initial velocity of the projectiles is $v = 1.5 \times 10^8$ cm s^{-1}.

The electronic stopping cross-section involves calculating the energy loss given by equation (4.12) along the path of the moving projectile in the $\langle 110 \rangle$ channel. In Figure 4.4 the $\langle 110 \rangle$ channel in silicon is shown together with a schematic view of the silicon atoms forming the 'rings' of this channel. The 'channeling' particle consecutively interacts, in this case, with the half rings AAA, BBB, etc. of this channel. The electronic energy loss in the interaction with AAA, BBB, etc., will be three times that of a single binary collision. For boron ($Z_1 = 5$) the energy loss will be

$$-\Delta E_{AAA} = 3 \times \frac{4.3 \times 10^{-8}(Z_1 + Z_2)^{5/3}v}{[1 + 0.31(Z_1 + Z_2)^{1/3}R_0)^5]} = 18.78 \text{ eV}, \qquad (4.17)$$

where $R_0 \approx 2.036$ Å for the $\langle 110 \rangle$ channel in silicon.

Taking into account the fact that these 'rings' are lying at distances of ≈ 1.92 Å from each other, we can calculate their average contribution per unit length to the electron energy loss:

$$-\Delta E_{\langle 110 \rangle} = -\Delta E_{AAA}/1.92 \text{ Å } = 9.79 \text{ eV Å}^{-1}.$$

In terms of the effective cross-section this result becomes

$$S_{\langle 110 \rangle} = \frac{\Delta E_{\langle 110 \rangle}}{N_0} \quad \frac{\text{eV Å}^2}{\text{atom}}$$

or

$$S_{\langle 110 \rangle} = 10^{-2}\frac{\Delta E_{\langle 110 \rangle}}{N_0} \quad \frac{\text{eV cm}^2}{\text{atom}} \times 10^{-14},$$

where $N_0 = 0.05$ atoms/Å^3 is the bulk density of silicon. In Figure 4.5 we compare the calculated electronic cross-section with the measured one by Eisen (1968) for the $\langle 110 \rangle$ channel in crystalline silicon.

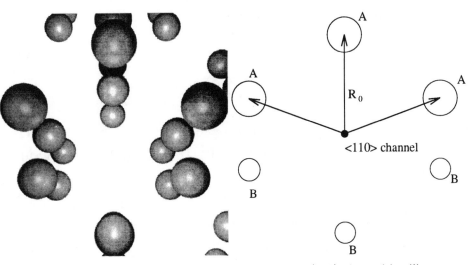

Fig. 4.4 Stereographic and schematic views of the ⟨110⟩ channel in silicon.

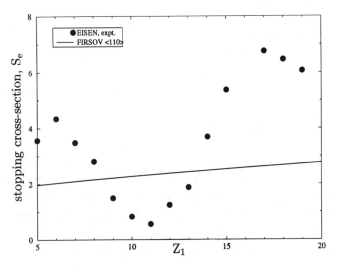

Fig. 4.5 The Z_1-dependence of the electronic stopping cross-section ($v = 1.5 \times 10^8$ cm s^{-1}) for the ⟨110⟩ direction in crystalline silicon. Experimental points are from Eisen (1968).

4.2.2 *Lindhard and Scharff electronic stopping (non-local)*

Instead of considering, as in Firsov theory, two isolated atoms colliding with their undistorted electronic shells, Lindhard and Scharff (1961) used a model of a slow heavy ion moving in a uniform electron gas. Electrons, impinging on the ion, transfer net energy which is proportional to their drift velocity (relative to the ion). As in the geometric model of Firsov, Lindhard and Scharff found that the electronic energy loss was proportional to the velocity:

$$S_e(E) = \xi_e 8\pi e^2 a_B \frac{Z_1 Z_2}{Z} \frac{v}{v_B} \frac{\text{eV Å}^2}{\text{atom}}, \tag{4.18}$$

with $Z = (Z_1^{2/3} + Z_2^{2/3})^{3/2}$, $a_B = \hbar^2/(me^2) = 0.529$ Å is the Bohr radius, $\xi_e \approx Z_1^{1/6}$ and (v/v_B) (a.u.) $= 6.343 \times 10^{-3} [E\,(\text{eV})/M_1]^{\frac{1}{2}}$. As before the units of M_1 are a.m.u.

4.2.3 Oen and Robinson electronic stopping

The two inelastic loss models described in Sections 4.2.1 and 4.2.2 differ in the way in which they treat the electronic energy losses. Firsov's model is a binary local model, which depends strongly on how closely two nucleii approach each other. On the other hand, the model of Lindhard and Scharff assumes that the inelastic energy loss of a particle depends on its energy and the path length it traverses.

Oen and Robinson (1976) developed a model that can be used either locally or non-locally. They start with a local inelastic loss model, which decays exponentially as a function of the distance of closest approach

$$-\Delta E(p, E) = K E^{1/2} \frac{\gamma^2}{2\pi a^2} e^{-\gamma R_0(p,E)/a},$$

where a is the screening length and γ is a parameter taken originally as 0.3 to connect the above equation with the Molière potential. However, γ can be used as a free parameter to make the inelastic energy loss follow the electron density in the target atom. The constant K can be determined as in any non-local model (Robinson 1993).

The Oen and Robinson electronic stopping cross-section is

$$S_e^{OR} = 2\pi \int_0^\infty p\, \Delta E(p, E)\, dp = K E^{1/2} \sigma(\varepsilon).$$

The deflection factor is

$$\sigma = \int_0^\infty \xi e^{-(\gamma/a) R_0(a\xi/\gamma, \varepsilon)}\, d\xi,$$

where

$$\varepsilon = E \frac{m_2}{m_1 + m_2} \frac{a}{Z_1 Z_2 e^2},$$

where the $4\pi\varepsilon_0$ factor has been omitted here and subsequently in this chapter.

At high energies $R_0(p, E) \approx p$, $\sigma \approx 1$ and $S_e^{OR} \approx KE^{1/2}$. For low energies $\sigma < 1$ and its effect is to reduce the electronic energy loss at low projectile velocities in a plausible manner (Robinson, 1993).

4.3 Z_1 oscillations in the electronic stopping

In Figure 4.5 it can be seen that Firsov's formula does not represent the oscillatory behaviour of the electronic cross-section with Z_1. This Z_1-dependence is enhanced under channelling conditions, under which the moving particles experience glancing collisions with the lattice atoms at an almost constant impact parameter. Several authors (Cheshire, Dearnley and Poate, 1968, Bhalla and Bradford, 1968, El-hoshy and Gibbons, 1968, Winterbon, 1968) suggested modifications to the existing theories of Firsov and Lindhard. These authors, considering shell effects in the electronic structure, introduced *effective nuclear charge*, which gave oscillatory behaviour to the calculated electronic stopping, similar to that observed in the experiment. The basic criticism of the Firsov model was the use of the semi-classical Thomas–Fermi model of the atom. Briggs and Pathak (1973, 1974) later developed a model in which the electronic stopping of heavy atoms moving through an electron gas was due to the elastic scattering of the electrons by the moving atom. They concluded that the Z_1 oscillations 'were clearly *not* dependent upon the inclusion of shell effects into the atomic potential'. They clearly demonstrated that, even using a Thomas–Fermi potential, the momentum transfer cross-section was strongly Z_1-dependent, having minima and maxima very similar to the variation seen in the experimental stopping power and concluded that 'the Z_1 oscillations are a fundamental feature of low-velocity stopping power under conditions where close atomic collisions are rare'.

A similar approach was used by Echenique *et al.* (1986). The electronic energy loss presented in the papers by Briggs and Pathak and Echenique *et al.* has the following form

$$-\frac{dE}{dx} \propto n v v_F \sigma_{tr}(v_F), \tag{4.19}$$

where

$$\sigma_{tr} = \frac{4\pi}{k_F^2} \sum_{l=0}^{\infty} (l+1) \sin^2(\eta_l - \eta_{l+1}) \tag{4.20}$$

is the *momentum-transfer cross-section*, k_F^2 (Ryd) is the electron energy in the centre-of-mass co-ordinate system, v_F is the Fermi velocity and η_l are the phase shifts induced by the field of the atom in the lth partial wave in the expansion of the electron wavefunction[1]. The stopping power for slow ions could then be rewritten in terms of scattering theories as

$$-\frac{dE}{dx} = \frac{3v}{k_F r_s^3} \frac{\hbar}{m} \sum_{l=0}^{\infty} (l+1) \sin^2 (\eta_l - \eta_{l+1}), \tag{4.21}$$

where $r_s = [3/(4\pi n)]^{1/3}$ is a measure of the electron density defined as the radius of a sphere whose volume is equal to the volume per conduction electron. The final expression for the stopping power depends on the approximation used to calculate the momentum-transfer cross-section. Echenique and co-workers defined in operational manner an effective charge

$$Z_1^* = \left[\left(\frac{dE}{dx} \right)_{Z_1 > 1} \middle/ \left(\frac{dE}{dx} \right)_{\text{proton}} \right]^{1/2}. \tag{4.22}$$

Using the density-functional approach and the above way of defining an 'effective charge', they succeeded in scaling the electronic stopping powers for $Z_1 \geq 2$ particles to that of protons. A complete discussion of the problem can be found in Echenique, Flores and Ritchie (1990).

Most of the theories of electronic stopping at low energies use the local density approximation. In this approximation, it is assumed that each volume element in the solid is an independent free plasma of electron density $n(r)$. Then, the stopping power, S_e, is position-dependent, proportional to the velocity of the ion. The existing theories give different expressions for S_e, depending on the way in which the plasma in the solid is described: as an ideal degenerate gas (Pathak and Youssoff, 1971), a Fermi liquid (Trubnikov and Yavlinskiĭ, 1965), a Hartree–Fock type quantum gas (Kitagawa and Ohtsuki, 1974) or an electronic system treated with a density-functional formalism (Echenique et al., 1981).

In the linear-response theory, the stopping power of a moving charged particle is given by

$$\frac{dE}{dx} = \frac{4Z_1^2 e^4 \hbar}{\pi v} \int_0^{\infty} \omega \, d\omega \int d\boldsymbol{q} \, \text{Im} \, K^{\text{r}}(\boldsymbol{q}, \omega)$$

$$\times \frac{\delta(\hbar \boldsymbol{q} \cdot \boldsymbol{v} + \hbar^2 q^2/(2m_1) + \hbar \omega)}{q^4}, \tag{4.23}$$

where m_1, v and Z_1 are the mass, velocity and atomic number of the incident ion respectively; $\hbar \boldsymbol{q}$ is the momentum transfer, and $K^{\text{r}}(\boldsymbol{q}, \omega)$ is the Fourier

---

[1] The phase shifts can be calculated by numerically solving the radial part of the Schrödinger equation (Briggs and Pathak, 1974). Detailed analysis can be found in Massey and Burhop (1969) chapter 6.

component of the retarded Green function $G^r(q, t)$ over time t, which usually appears in the calculation of the dielectric function ϵ^{-1}. $G^r(q, t)$ represents the density–density correlation of the system.

Integrating over the angular part of q, Kitagawa and Ohtsuki (1974) succeeded in separating the v-dependence of dE/dx

$$-\frac{dE}{dx} = -\frac{i4Z_1^2 e^4}{\pi} \int \frac{dq}{q} \int_{-\infty}^{\infty} \operatorname{Im} K^r(q, \omega)\, d\omega$$

$$\times \int_{-\infty}^{\infty} dt \exp\left(-i\omega t\right) j_1(qvt), \tag{4.24}$$

where $j_1(qvt)$ is the spherical Bessel function which, for the low-velocity case, can be written

$$j_1(qvt) \approx \frac{1}{3} qvt.$$

On replacing the above approximation for the Bessel function into equation (4.24) we obtain

$$-\frac{dE}{dx} = -\frac{8Z_1^2 e^4}{3} v \int dq \frac{d}{d\omega} \left[\operatorname{Im} K^r(q, \omega)\right]\Big|_{\omega=0}, \tag{4.25}$$

which shows that, in the low-velocity regime, the stopping power is proportional to the velocity.

For the Fourier component, $K^r(q, \omega)$, of the retarded Green function, Kitagawa and Ohtsuki obtained

$$\frac{d}{d\omega} \left[\operatorname{Im} K^r(q, \omega)\right]\Big|_{\omega=0} = -\frac{m^{*2}}{2\pi\hbar} \frac{q^3}{[q^2 + q_T^2 f(q)]^2}, \tag{4.26}$$

where m^* is the effective mass of the electron and

$$f(q) = 1 - \frac{q^2}{2(q^2 + q_F^2)}$$

with q_F the Fermi momentum and

$$q_T^2 = 4e^2 m^2 q_F / (\pi\hbar^2).$$

Combining equations (4.25) and (4.26) we obtain for the stopping power

$$-\frac{dE}{dx} = \frac{2Z_1^2 e^4 m^{*2}}{3\pi\hbar^3} v \int_0^1 d\zeta \frac{\zeta}{[\zeta + m^*/m\alpha f(\zeta)]^2}, \tag{4.27}$$

where $\alpha = e^2/(\pi\hbar v_F)$, $\zeta = q^2/(4q_F^2)$ and $f(\zeta) = (2\zeta + 1)/(4\zeta + 1)$.

Now we consider the two limiting cases: a high-density electron gas, $\alpha \ll 1$, and a low-density electron gas with $\alpha \gg 1$. For the case of $\alpha \ll 1$, $m^* \approx m$ and we obtain the Fermi–Teller formula (Fermi and Teller, 1947)

$$-\frac{dE}{dx} \simeq \frac{2Z_1^2 e^4 m^2}{3\pi\hbar} v \ln\left(\frac{\hbar v_F}{e^2}\right), \tag{4.28}$$

indicating that, for a high-density plasma, the correlation between electrons of the system can be neglected.

For a low-density electron plasma ($\alpha \gg 1$), $(m^*/m)\alpha \approx \frac{3}{4}$, $m^* \approx 3\pi q_F \hbar^2/(4e^2)$, and the stopping power obtained by integrating (4.27) is

$$-\frac{dE}{dx} \simeq 0.32\pi^{7/3} Z_1^2 \hbar v n^{2/3}, \tag{4.29}$$

where $q_F = (3\pi^2 n)^{1/3}$ and n is the electron density.

The stopping power for slow ions can be calculated using the momentum-transfer cross-section σ_{tr}. On the basis of non-linear calculations of electron-density fluctuations, which give a result quite different from those of linear-response theory (equations (4.24) and (4.25)), Echenique *et al.* (1981) evaluated the phase shifts, η_l, within the density-functional formalism. Figure 4.6 shows the stopping powers as a function of electron density calculated with the linear-response theory, the density-functional formalism and two approximations, the Fermi–Teller and Kitagawa–Ohtsuki ones, to the linear theory. We note two important points. First, most metals exhibit $\alpha \approx 1$ and the two extreme cases, Fermi–Teller and Kitagawa–Ohtsuki, are not applicable. Secondly, in the calculation of the stopping power the overall electron density is used and not that of valence electrons only, as is the case for high energies.

4.4 Z_2 oscillations in the electronic stopping

When the electronic stopping for channelled ions is plotted versus the atomic number of the target material, Z_2, it shows oscillations similar to those for Z_1. Pathak (1974) using the same theory discussed above, concluded that the variation in the electronic stopping power with Z_2 under channelling conditions is not of the same nature as the Z_1 oscillations and that the Z_2-dependence primarily comes from the difference in the 'effective density' of the target electrons. Brandt (1982), having examined experimental data, concluded that a comprehensive description of low-velocity electronic stopping powers could be given, if reference is made not to Z_2 but to the valence electron density.

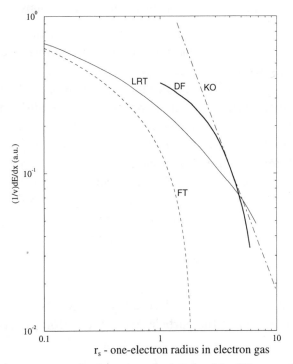

Fig. 4.6 Stopping powers of a proton calculated with different theories: density-functional formalism (DF); linear-response theory (LRT); the Fermi–Teller formula (FT) and the Kitagawa–Ohtsuki approximation for low density (KO). In all cases $v \ll v_F$.

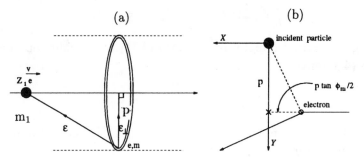

Fig. 4.7 (a) The Coulomb field $\mathscr{E}_\perp$ acting on an electron of mass m and charge e due to the passage of a heavy particle of mass m_1 and charge Z_1e with velocity v and impact parameter p. (b) The centre of mass (CM) reference frame where the heavy incident particle is practically at rest. Here $\phi_m/2$ is the scattering angle defined in Chapter 2.

4.5 Electronic stopping for high-energy ions

At high energies, range III in Figure 4.2, the heavy particles of charge Z_1e, during their passage through matter, lose energy almost entirely through inelastic collisions with the bond electrons in the atoms of the material. To calculate the energy loss we consider the electrons as free and at rest. The trajectory of the heavy particle is not appreciably affected by the light electrons and the collision lasts such a short time that the electron acquires an impulse without changing its position. To find out the energy transfer during the collision, we need to calculate the momentum transfer to the electron, caused by the electric field of the incident particle, at the position of the electron. Only the transverse electric field $\mathscr{E}_\perp$ has a non-vanishing time integral. Consequently, the impulse ΔP is perpendicular to the trajectory of the incident particle, Figure 4.7(a), and has magnitude

$$\Delta P = \int e\mathscr{E}_\perp \, dt = \int_{-\infty}^{\infty} e\mathscr{E}_\perp \frac{dx}{v}, \tag{4.30}$$

where $\mathscr{E}_\perp$ is the component of the electric field at the position of the electron normal to the trajectory of the particle. We can transform equation (4.30) to the form

$$\int e\mathscr{E}_\perp \frac{dx}{v} = \frac{e}{2\pi p v} \int_S \mathscr{E}_\perp 2\pi p \, dx, \tag{4.31}$$

S being the surface of the cylinder. The last integral is the flux of $\mathscr{E}$ through the cylinder having the trajectory as its axis. By applying Gauss's theorem to the last integral, it becomes

$$\int_S \mathscr{E}_\perp 2\pi p \, dx = \int_S \mathscr{E}_\perp \, dS = 4\pi Z_1 e. \tag{4.32}$$

Replacing the second integral in equation (4.31) by its value obtained from equation (4.32) gives

$$\Delta P = \frac{2Z_1 e^2}{p v}. \tag{4.33}$$

The velocity of the moving particle is taken to be constant during the collision. The energy transferred to the electron is

$$\Delta E(p) = \frac{(\Delta P)^2}{2m} = \frac{2Z_1^2 e^4}{m v^2} \left(\frac{1}{p^2} \right). \tag{4.34}$$

The scattering angle in the CM co-ordinate system is given by

$$2\tan\left(\frac{\phi_m}{2} \right) = \frac{Z_1 e^2}{P v p} \simeq \left. \frac{\Delta P}{2P} \right|_{\text{if } \Delta P \ll P}. \tag{4.35}$$

Because of the mass difference between the heavy incident particle and the electron, the incident particle is practically at rest in the CM coordinate system. Then, equation (4.35) holds with $P = mv$ as the momentum in that frame. From Figure 4.7(b) it can be seen that the square of the minimum distance between the incident particle and the electron is $p^2 + (p \tan \phi_m/2)^2 = p^2 + \left[Z_1 e^2/(mv^2) \right]^2$. Let us call the second term $p_{\min}$ and substitute for the impact parameter in equation (4.34) the distance estimated above:

$$\Delta E(p) \simeq \frac{(\Delta P)^2}{2m} = \frac{2 Z_1^2 e^4}{mv^2} \left(\frac{1}{p^2 + p_{\min}^2} \right). \tag{4.36}$$

A fast particle passing through matter 'sees' electrons at various distances from its path. To find the energy loss of the incident particle per unit distance, we multiply the differential cross-section $d\sigma_S$, as defined in equation (2.58) by the energy transfer $\Delta E(p)$ and integrate over all impact parameters. Thus the energy loss is

$$-\frac{dE}{dx} = n \int d\sigma_S \, \Delta E(p) = 2\pi n \int \Delta E(p) \, p \, dp$$

$$= 4\pi n \frac{Z_1^2 e^4}{mv^2} \int_{p_{\min}}^{p_{\max}} \frac{dp}{p} = 4\pi \frac{Z_1^2 e^4}{mv^2} N_0 Z_2 \times L, \tag{4.37}$$

where

$$L = \int_{p_{\min}}^{p_{\max}} \frac{dp}{p} = \ln \left(\frac{p_{\max}}{p_{\min}} \right)$$

and $n = N_0 Z_2$ is the electron density in the stopping material.

For very distant collisions the approximate result given by equation (4.34) is in error because of the binding of the electrons in atomic orbits. Thus, to estimate $p_{\max}$, we consider that the electrons are not free but are *bound in atomic orbits*. Because $p_{\max}$ is an upper limit defining the distant collision, the Coulomb field acting on the electron can be regarded as a perturbation. The duration of such a perturbation is the time $\tau \propto p/v$ during which the heavy particle is near the electron. If τ is large enough compared with $1/\omega_i$, we have adiabatic behaviour of the electron i.e. the electron will make many cycles around the atom as the incident particle slowly passes by and will be adiabatically influenced by the fields with no net transfer of energy. The dividing point comes at some predefined cut-off impact parameter $p_{\max}$, at which the collision time and the orbital period of the electron are comparable:

$$p_{\max} = \frac{v}{\omega},$$

where ω is some average oscillation frequency[1] of the electron in the atom. The precise result is

$$p_{max} = \frac{2}{C}\frac{v}{\omega} \, ,$$

where $2/C = 2e^{-\gamma} = 1.123$, $\gamma = 0.577$ being Euler's constant. A detailed discussion on the correction, $2/C$, was given by Bloch (1933). For impact parameters greater than p_{max} it can be expected that the energy transfer falls below that defined by equation (4.34), going rapidly to zero for $p \gg p_{max}$.

The classical limit for p_{min} depends on the maximum energy which could be transferred between the heavy particle and the electron in a collision. We can obtain the lower limit on the impact parameter p_{min} for which our approximate calculation is valid, by equating equation (4.34) to the maximum allowable energy transfer (cf. equation (2.42))

$$\Delta E(p_{min}) = T_m = \frac{4m_1 m}{(m_1 + m)^2}\frac{1}{2}m_1 v^2 \approx 2mv^2 . \qquad (4.38)$$

This implies a minimum *classical* impact parameter

$$p_{min}^{cl} = \frac{Z_1 e^2}{mv^2}$$

below which our approximate result, equation (4.34), must be replaced by equation (4.36), which tends to equation (4.38) as $p \to 0$.

Introducing these values for p_{max} and p_{min} in L, equation (4.37) gives Bohr's (1913, 1915) classical stopping formula

$$-\frac{dE}{dx} = 4\pi \frac{Z_1^2 e^4}{mv^2} N_0 Z_2 \ln\left(\frac{1.123 mv^3}{Z_1 e^2 \omega}\right). \qquad (4.39)$$

Bohr's formula gives a reasonable description of the energy loss of relatively slow alpha particles and heavier nuclei. However, for electrons, protons or swift alpha particles, it considerably overestimates the energy loss. The reason is that, for lighter particles, the quantum mechanical nature of the processes causes a breakdown of the classical result. There are two important quantum effects that should be considered; the discreteness of the possible energy transfers and limitations due to the wave nature of the particles and the uncertainty principle.

----

[1] If f_i is the oscillator strength for frequency ω_i and $Z_2 = \sum_i f_i$, then

$$\ln \omega = \frac{1}{Z_2}\sum_i f_i \ln \omega_i .$$

We can illustrate the discrete nature of the energy transfer by calculating the classical energy transfer equation (4.34) at $p \approx p_{\max} = v/\omega$. At such impact parameters, this is roughly the smallest energy transfer that is of importance in the energy loss process. Assuming, for simplicity, only the ground frequency, ω_0, and keeping in mind that $v_B = e^2/\hbar$, $a_B = \hbar^2/(me^2)$ and $\omega_0 = v_B/a_B = me^4/\hbar^3$, we find

$$\Delta E(p_{\max}) \approx \frac{2Z_1^2 e^4}{mv^2}\frac{1}{p_{\max}^2} = 2Z_1^2\left(\frac{v_B}{v}\right)^4\frac{\hbar^3\omega_0}{me^4}\hbar\omega_0 = 2Z_1^2\left(\frac{v_B}{v}\right)^4\hbar\omega_0. \quad (4.40)$$

We see that for swift particles ($v \gg v_B$) the classical energy transfer, equation (4.40), is very small compared with the smallest excitation energy in the atom, $\hbar\omega_0$. Obviously, the energy transfer cannot be smaller than the minimum allowed by quantum mechanics. That is why we have to re-interpret the classical result and apply it in a statistical sense such that energy transfer does not occur in every collision. In fact, it is a rather rare event whereby in a few collisions an appreciable excitation occurs, yielding a small average value over many collisions. In this statistical sense, the quantum mechanism for discrete energy transfers and the classical process with a continuum of possible energy transfers can be reconciled.

Another extreme value is the minimum impact parameter, $p_{\min}$. From the uncertainty principle we know that the path of the electron can be defined only to within uncertainty $\Delta x \geq \hbar/P$, the de Broglie wavelength of the electron. For impact parameters, p, less than this uncertainty, classical concepts fail. Because of the wave nature of the electron and its uncertainty in the space within length Δx, for impact parameters $p < \Delta x$ the exact quantum-mechanical calculation of the energy transfer will be smaller than the classical result given by equation (4.34). Thus $\Delta x \propto \hbar/P$ is the quantum analogue for the minimum impact parameter (cf. equation (4.38)). For a heavy incident particle colliding with the electron, the momentum of the electron, in the co-ordinate frame in which the incident particle is at rest, is $P = mv$. Therefore, the quantum-mechanical minimum impact parameter is

$$p_{\min}^{\mathrm{qm}} = \frac{\hbar}{mv}.$$

This minimum impact parameter introduced in L gives the stopping formula

$$-\frac{dE}{dx} = 4\pi\frac{Z_1^2 e^4}{mv^2}N_0 Z_2 \ln\left(\frac{mv^2}{\hbar\omega}\right).$$

A more precise calculation of the stopping power, based on Born's approximation, has been performed by Bethe[1] (1930)

$$-\frac{\mathrm{d}E}{\mathrm{d}x} = 4\pi \frac{Z_1^2 e^4}{mv^2} N_0 Z_2 \ln\left(\frac{2mv^2}{I}\right). \tag{4.41}$$

Here I is the mean excitation energy for the atom in its ground state. The quantity I is difficult to estimate exactly and the interested reader is referred to Bethe's original paper.

Use of Born's approximation requires that the amplitude of the wave scattered by the field of the atomic electron be small compared with the amplitude of the undisturbed incident wave. The criterion for this is that (Williams 1945)

$$1 \gg \frac{2Z_1 e^2}{\hbar v} = \kappa.$$

In 1933, by approximating the perturbation of the wavefunctions of the atomic electrons due to the incident particle, Bloch combined Bohr's and Bethe's formulae. The *classical* Thomson (1912) cross-section gives

$$L_{\text{class}} = \int_0^{p_{\max}=\frac{2v}{C\omega}} \frac{p\,\mathrm{d}p}{p^2 + p_{\min}^2}, \quad p_{\min} = \frac{Z_1 e^2}{mv^2} = \frac{Z_1 e^2}{\hbar v}\frac{\hbar}{mv} = \frac{\kappa}{2}\frac{\hbar}{mv}.$$

Introducing an angular momentum

$$p = \frac{1}{mv} \times (\text{ang.momentum}) = \frac{\hbar}{mv}l$$

L_{class} becomes

$$L_{\text{class}} = \int_0^{l_{\max}} \frac{l\,\mathrm{d}l}{l^2 + \kappa^2/4}, \quad l_{\max} = \frac{2mv^2}{C\hbar\omega}.$$

Scattering theory shows that l is an integer value. Then we obtain

$$L_{\text{Bloch}}(\kappa) = \sum_{l=0}^{l_{\max}} \frac{l+1}{(l+1)^2 + \kappa^2/4}.$$

If $\kappa = 0$

$$L_{\text{Bloch}}(0) = L_{\text{Bethe}} = \sum_{l=0}^{l_{\max}} \frac{1}{l+1} \approx \ln\left(\frac{2mv^2}{C\hbar\omega}\right) + \ln C,$$

where one could neglect $\ln C$ for high energies.

---

[1] In his theory, the differential cross-section for the process of energy transfer between the moving particle and the electron is essentially given by the square of the matrix element of the Coulomb interaction between appropriate initial and final states. Plane waves have been used for the wavefunctions for the incident and 'scattered' heavy particle.

If κ is large

$$L_{\text{Bloch}}(\kappa \gg 1) = L_{\text{Bohr}} \simeq \ln\left(\frac{2mv^2}{C\hbar\omega}\right) - \ln\frac{\kappa}{2}.$$

For small κ

$$L_{\text{Bloch}} = L_{\text{Bethe}} - \frac{\kappa^2}{4}\sum_{l=0}^{l_{\max}}\frac{1}{(l+1)^3} + \cdots$$

and therefore the Bloch correction is proportional to $[Z_1e^2/(\hbar v)]^2 = \kappa^2/4$.

4.6 Ion–electron interaction with regard to MD simulations

Molecular dynamics (MD) simulations in metals are now usually carried out by applying effective many-body potentials, like the embedded-atom model (EAM) or the Finnis–Sinclair model (FSM) (see Chapter 3). These potentials give a quite reliable description of crystalline structure, elastic constants, phonon dispersion relations, defect properties and dynamic correlation in the liquid phase. However, good potentials and large crystal sizes do not necessarily lead to a good description of radiation effects in crystals. The simulation of, for example, thermal spikes, radiation defects and annealing is not necessarily close to reality if the electronic energy loss and description of the thermal conductivity are not taken into consideration. Both of the above mechanisms are responsible for the way energy is dissipated into the crystal. The electronic energy loss covers an energy range from very high energies down to a few electronvolts whereas the thermal conductivity, caused by electron–phonon interactions, might become dominant in the energy interval of a few electronvolts down to 10^{-2} eV.

Let us first consider an ion–electron interaction. Because, at present, large-scale MD simulations are performed for energies no more than a few times 10^4 eV, the electronic stopping is confined to region I of Figure 4.2, where the interactions between electrons and ions are elastic and the stopping is treated as a retarding force proportional to the velocity of the ion. Two different electronic loss mechanisms can be incorporated into the Newton equations – a continuous loss mechanism or one that depends on the local electron density. As was shown previously, in equations (4.15), (4.16) and (4.18), the electronic stopping cross-section is of the form

$$S_e(E) = KE^{1/2}$$

and the stopping power

$$\frac{dE}{dx} = -N_0 S_e(E) = -N_0 K E^{1/2}.$$

Since

$$\frac{dE}{dx} = \frac{1}{v}\frac{dE}{dt},$$

$$\frac{dE}{dt} = -N_0 K v E^{1/2} = -N_0 K \left(\frac{m_2}{2}\right)^{-1/2} E = -\frac{1}{\tau_v} E, \qquad (4.42)$$

where

$$\tau_v = \frac{1}{N_0 K}\left(\frac{m_2}{2}\right)^{1/2}$$

can be considered as the lifetime of the kinetic energy due to electronic losses.

If we make the assumption that the retarding force due to electronic energy loss is proportional to the velocity then

$$m_2 \ddot{R} = F - \beta \dot{R},$$

where β is the constant of proportionality which, from equation (4.42), is given by $\beta = m_2/\tau_v$.

In the other approach, the local electron density model, it can be demonstrated using Firsov's model for electron energy loss, equation (4.4), that

$$\Delta F_b = m(\dot{R}_a - \dot{R}_b)\int_S (nu/4)\,dS,$$

where R_a is the position vector of atom a and R_b that of atom b and $\Delta F_a = -\Delta F_b$. The evaluation of the above integral is a long task. Beeler (1983) gave a solution using a technique due to Wederpohl (1967):

$$\Delta F_b = (\dot{R}_a - \dot{R}_b)(8.06172 \times 10^{-2} Z^2)\left(e^{-\mathscr{C}}\sum_{m=0}^{7}(\mathscr{C}^m/m!)\right)\frac{eV}{\text{Å}},$$

where the velocities are expressed in Å per 10^{-14} s. The quantity $\mathscr{C}$ is given by

$$\mathscr{C} = 13.4646 Z^{1/12}(|R_a - R_b|)^{1/4},$$

where Z is the atomic number of the interacting particles (assumed to be of the same type) and the positions $\dot{R}_a$ and $\dot{R}_b$ are expressed in ångström units.

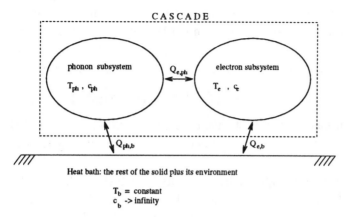

Fig. 4.8 The thermodynamic model for energy transfer in the cascades.

Similar results are obtained (Eltekov *et al.*, 1970) using the same electronic energy loss model:

$$\Delta F_{\rm b} = (\dot{R}_{\rm a} - \dot{R}_{\rm b})\, \frac{0.7\hbar}{\pi a^2}$$

$$\times \left(\frac{Z_1^2}{(1 + 0.8 q Z_1^{1/3} R_0/r)^4} + \frac{Z_2^2}{[1 + 0.8(1 - q) Z_2^{1/3} R_0/r]^4} \right),$$

where $r = 0.47$, $q = [1 + (Z_2/Z_1)^{1/6}]^{-1}$ and $R_0 = |R_{\rm a} - R_{\rm b}|$.

The ion–electron interaction does not cover the whole possible range in which the energy of the cascade dissipates. Let us regard the solid as a three-component system, with each characterised by its temperature, and look at the energy exchange between the components, Figure 4.8. The three components are (i) the cascade nuclei (plus core electrons moving rigidly with them) which have temperature $T_{\rm ph}$ and specific heat $c_{\rm ph}$; (ii) the electrons, which are dynamically independent in the extent of the cascade and which have temperature $T_{\rm e}$ and specific heat $c_{\rm e}$; (iii) the rest of the solid, which stays at a constant ambient temperature $T_{\rm b}$ and, like all ideal heat baths, has infinite specific heat.

We are looking for a simple description of the thermal behaviour of the cascade once energy has been deposited during the collision phase. This behaviour depends both on specific heats and on electron–phonon coupling (i.e. energy transfer between systems (i) and (ii)). If there is no electron–phonon coupling, the hot ion temperature decays with a characteristic time constant $\tau_{\rm ph}^{-1} = Q_{\rm ph,b}/c_{\rm ph}$ and the electron temperature recovers with time constant $\tau_{\rm e}^{-1} = Q_{\rm e,b}/c_{\rm e}$. Because electrons dominate heat conduction in metals, the heat transfer coefficient $Q_{\rm e,b}$ will usually exceed $Q_{\rm ph,b}$. When there is energy transfer between the two cascade subsystems we define $\sigma = c_{ph}/c_e$ ($\sigma \ll 1$ because c_e is small for T_e much less than the Fermi temperature).

For very rapid transfer of energy between these two subsystems, the time τ characterising their common trend to ambient temperature, is

$$\tau^{-1} = (\tau_e^{-1} + \sigma\tau_{ph}^{-1})(1 + \sigma).$$

For a Fermi gas of free electrons, c_e is very small (σ large) and the characteristic time τ is essentially τ_{ph}. Therefore, for a rapid transfer we can apply simple models.

Another ratio of importance concerns how fast equilibrium between the lattice and the electron systems is reached, relative to the rate at which these combined systems fall into equilibrium with the heat bath

$$\frac{\tau_{e,ph}^{-1}}{\tau^{-1}} = \frac{Q_{e,ph}}{Q_{e,b}Q_{ph,b}}\frac{(1+\sigma)^2}{\sigma}.$$

When one of the specific heats is much larger than the other ($\sigma \ll 1$ or $\sigma \gg 1$), reaching equilibrium between electrons and ions is much faster than their joint recovery. This is obvious, since the system with the smaller capacity rapidly adopts the temperature of the other and the large heat capacity determines a slow return to the bath temperature. This model shows that there are two possible asymptotic regimes: electronic heat conduction dominant and electronic heat sink dominant.

Whether or not the ion motion is affected by the electrons in metals depends on the effectiveness of the electrons as an energy sink, their effectiveness in the conduction of heat and the extent to which electrons can cause other behaviour. We can conclude from simple solid-state arguments that (i) electrons are a poor energy sink, with a very low specific heat as long as Fermi statistics apply, (ii) electrons are a very good means of transferring heat in metals and (iii) electron and ion subsystems reach equilibrium separately, faster than they exchange energy, partly justifying the assumptions of well-defined ion and electron temperatures in the thermodynamic model.

To incorporate the simple model discussed above into a MD simulation we need to know the rate of energy transfer to the electrons as a function of R and $\dot{R}$, which are not necessarily related to T_{ph} or to the electronic thermal conductivity as a function of T_e. The physical picture of the processes involved is as follows. Electrons and phonons initially in thermal equilibrium are perturbed by a primary knock-on atom (PKA). The model of the coupling between the phonons and the electrons should reproduce the electron–phonon coupling when in equilibrium and the electronic stopping power when out of equilibrium. One can consider the electrons as a perfect heat sink (σ is large), i.e., having infinite thermal conductivity compared with ions; hence, the trend to equilibrium will depend entirely on the ion subsystem ($\tau \propto \tau_{ph}$).

The interaction of the lattice atoms in MD is deterministic, involving integration of the equation of motion. The temperature of the simulation can be incorporated via stochastic forces. Then, the detailed interaction of the atoms with the heat bath is neglected and only taken into account by the stochastic force. We can use classical stochastic Langevin equations, which describe the interactions of an ensemble of classical particles with an irreversible thermal reservoir. In addition, the atoms interact with each other via some deterministic force. The Langevin equations of motion are given by

$$m_2 \ddot{\boldsymbol{R}} = \boldsymbol{F} + \vec{\mathscr{R}}(t) - \beta \dot{\boldsymbol{R}}. \tag{4.43}$$

The right-hand side represents the coupling to the heat bath. Algorithms for solving this equation are given in Chapter 8. Here we describe how the constant β and the random force $\vec{\mathscr{R}}(t)$ can be determined.

The effect of the random force $\vec{\mathscr{R}}(t)$ is to heat the particle, whereas the other term, $\beta \dot{\boldsymbol{R}}$, represents the retarding force due to interaction between subsystems (i) and (ii) in our model and $\beta \propto \tau^{-1}$ is a constant measuring the strength of coupling to the thermal bath. The random force, $\vec{\mathscr{R}}(t)$, is normally distributed and defined by

$$\langle \vec{\mathscr{R}}(t) \rangle = 0,$$
$$\langle \vec{\mathscr{R}}(t) \cdot \vec{\mathscr{R}}(t_0) \rangle = 2\beta k_{\mathrm{B}} T_{\mathrm{b}} \delta(t - t_0), \tag{4.44}$$
$$\mathscr{P}(\mathscr{R}) = (2\pi \langle \mathscr{R}^2 \rangle)^{-1/2} \exp\left[-\mathscr{R}^2/(2\langle \mathscr{R}^2 \rangle)\right].$$

Here k_{B} is Boltzmann's constant and the second equation means that, at two different times t and t_0, the random force is uncorrelated.

The correlation times of the velocity and the random force are

$$\tau_v = m_2/\beta, \qquad \tau_{\mathscr{R}} = \beta k_{\mathrm{B}} T_{\mathrm{b}} \langle \mathscr{R}^2 \rangle^{-1}. \tag{4.45}$$

It is reasonable to require that the correlation time for the random force be much smaller than that for the velocity, $\tau_{\mathscr{R}} \ll \tau_v$. The numerical algorithm for solving the equations of motion involves a finite step size in time, i.e.

$$\boldsymbol{R}(t) \rightarrow \boldsymbol{R}(t + \Delta t), \qquad \dot{\boldsymbol{R}}(t) \rightarrow \dot{\boldsymbol{R}}(t + \Delta t).$$

Let us assume that, during the time step Δt, the particle experiences a constant random force and that the correlation time $\tau_{\mathscr{R}}$ is equal to Δt. Before each integration step we choose a random force, $\mathscr{R}_{\mathrm{n}}$, from a Gaussian distribution with mean zero and variance $\langle \mathcal{R}^2 \rangle$ according to equation (4.44). We still have to determine $\langle \mathcal{R}^2 \rangle$, which can be done using equation (4.45) with correlation time $\tau_{\mathcal{R}} = \Delta t$

$$\langle \mathcal{R}^2 \rangle = \beta k_B T_b \, \Delta t^{-1}.$$

The coupling parameter β can be estimated by analysing Figure 4.6 and using, for the initial approximation the Fermi–Teller formula, (4.28):

$$-\frac{1}{v} \frac{dE}{dx} = \frac{2Z_1^2 e^4 m^2}{3\pi \hbar^3} \, \ln \left(\frac{\hbar v_F}{e^2} \right) = Z_1^2 A \ln \left(\frac{1}{\alpha'} \right).$$

For a spherical Fermi surface

$$v_F = \frac{\hbar q_F}{m^*(\approx m)}, \qquad q_F = (3\pi^2)^{1/3} n^{1/3}.$$

Then

$$\frac{1}{\alpha'} = \frac{(3\pi^2)^{1/3} \hbar^2}{me^2} \, n^{1/3}$$

and the Fermi–Teller formula is reduced to

$$-\frac{1}{v} \frac{dE}{dx} = AZ_1^2 \ln(Bn^{1/3}) \quad \text{i.e.} \quad -\frac{dE}{dx} = \beta v,$$

where

$$A = 2e^4 m^2/(3\pi \hbar^3) = 4.982 \times 10^{-16} \text{ eV s } \text{Å}^{-2},$$

$$B = (3\pi^2)^{1/3} \hbar^2/(me^2) = 1.6371 \text{ Å}.$$

In order to make a closer approximation to the density-functional calculations of Echenique, Nieminen and Ritchie (1981), Caro and Victoria (1989) introduced a modification to the above method of calculating β. They wrote

$$\beta = AZ^{*2} \ln \left(Bn^{1/3} + C \right) + b. \tag{4.46}$$

In this formula b is the boundary damping constant and Z^* and C are adjustable parameters.

These parameters, Z^* and C, have to be chosen to reproduce the high-energy stopping power (depending essentially on Z^*) and the electron–phonon coupling at room temperature (depending both on Z^* and on C). An example of the evaluation of these two parameters was given by Proennecke *et al.* (1991).

4.7 Summary

The local and non-local models of electronic energy loss can be easily incorporated into both binary collision and molecular dynamics computer programs. For a non-local model in a BC program ΔR can be calculated from a knowledge of the particle's velocity and the distance that it travels between collisions. It is a simple matter to subtract ΔE from the energy of the particle before it undergoes the next collision. The energy loss in the local model, however, needs to be subtracted at the point of collision. This gives a different value for the scattering angle (2.20) than if ΔE is subtracted before the collision. This point is discussed further in Chapter 7.

The well-known TRIM program uses a semi-empirical model of electronic stopping due to Ziegler, Biersack and Littmark (1985) whereas the MARLOWE program uses the model of Oen and Robinson (1976).

Both local and non-local electronic energy loss models have been incorporated into MD codes. The non-local models are easily coded provided that β can be easily determined. Thus the model of Caro and Victoria (1989) can be easily coded if the interatomic potential is one in which the electron density naturally occurs, such as the embedded atom or Finnis–Sinclair models. However, this is not especially convenient for use with the covalent potentials described in Chapter 3, which are not formulated explicitly in terms of this density.

References

Beeler, J. R. (1983). *Radiation Effects – Computer Experiments*, North-Holland, Amsterdam, 363.
Bethe, H. A. (1930). *Ann. Phys.* **5** 325.
Bhalla, C. P. and Bradford, J. N. (1968). *Phys. Lett.* A **27** 318.
Bloch, F. (1933). *Z. Phys.* **81** 363; *Ann. Phys.* **16** 285.
Bohr, N. (1913). *Phil. Mag.* **25** 10.
Bohr, N. (1915). *Phil. Mag.* **30** 581.
Brandt, W. (1982). *Nucl. Instrum. Meth.* **194** 13.
Briggs, J. S. and Pathak, A. P. (1973). *J. Phys.* C **6** L153.
Briggs, J. S. and Pathak, A. P. (1974). *J. Phys.* C **7** 1929.
Caro, A. and Victoria, M. (1989). *Phys. Rev.* A **40** 2287.
Cheshire, I. M., Dearnley, G. and Poate, J. M. (1968). *Phys. Lett.* **27** 304.
Echenique, P. M., Flores, F. and Ritchie, R. H. (1990). *Solid State Physics – Advances in Research and Applications* Vol. 43, Academic Press, New York, 229.
Echenique, P. M., Nieminen, R. M., Ashley, J. C. and Ritchie, R. H. (1986). *Phys. Rev.* A **33** 897.
Echenique, P. M., Nieminen, R. M. and Ritchie, R. H. (1981). *Solid St. Comun.* **37** 779.
Eisen, F. H. (1968). *Can. J. Phys.* **46** 561.

El-hoshy, A. H. and Gibbons, J. F. (1968). *Phys. Rev.* **173** 454.

Eltekov, V. A., Karpuzov, D. S., Martinenko, Yu. V. and Yurasova, V. E. (1970). *Atomic Collision Phenomena in Solids*, Ed. D. W. Palmer, M. W. Thompson and P. D. Townsend, North-Holland, Amsterdam, 657–62.

Fermi, E. and Teller, E. (1947). *Phys. Rev.* **72** 399.

Firsov, O. B. (1959). *Sov. Phys.: J. Exp. Theor. Phys.* **36** 1076.

Kitagawa, M. and Ohtsuki, Y. H. (1974). *Phys. Rev.* B **9** 4719.

Lindhard, J. and Scharff, M. (1961). *Phys. Rev.* **124** 128.

Massey, H. S. and Burhop, E. H. (1969). *Electronic and Ionic Impact Phenomena* Vol 1, Clarendon Press, Oxford.

Marich, N. H. (1950). *Proc. Cam. Phil. Soc.* **46** 356.

Oen, O. S. and Robinson, M. T. (1976). *Nucl. Instrum. Meth.* B **19/20** 101.

Pathak, A. P and Youssoff, M. (1971). *Phys. Rev.* B **3** 3701.

Pathak, A. P. (1974). *J. Phys.* C **7** 3239.

Proennecke, S., Caro, A., Victoria, M., Diaz de la Rubia, T. and Guinan, M. W. (1991). *J. Mater. Res.* **6** 483.

Robinson, M. T. and Torrens, I. M. (1974). *Phys. Rev.* B **9** 5008.

Robinson, M. T. (1993). *Mat. Fys. Med.* **43** 27.

Thomson, J. J. (1912). *Phil. Mag.* **23** 449.

Trubnikov, B. A. and Yavlinskiĭ, Y. (1965). *Sov. Phys.: J. Exp. Theor. Phys.* **21** 167.

Wederpohl, P. T. (1967). *Proc. Phys. Soc.* **92** 79.

Williams, E. J. (1945). *Rev. Mod. Phys.* **17** 217.

Winterbon, K. B. (1968). *Can. J. Phys.* **46** 2429.

Ziegler, J. F., Biersak, J. P. and Littmark, U. (1985). *The Stopping and Range of Ions in Solids*, Pergamon, New York, 107.

5

Transport models

5.1 The Boltzmann transport equation

5.1.1 Introduction

Atomic particles are both deflected and slowed down after scattering by a target atom. This process is fundamental to the study of the penetration of ions in solid targets. A typical ion–solid experiment would involve many ion trajectories comprising several scatterings. Computer models tackle the problem head-on by calculating entire collision cascades from a representative set of trajectories. These results can then be used to evaluate average values such as the mean penetration depth and the mean number of particles ejected within a certain angle or energy range. However, the computer models often contain details that are not accessible to experimental observation and vast amounts of computing time can often be expended in generating these average results.

Computational techniques are discussed in more detail elsewhere in this book. In this chapter a probabilistic description amenable to analytic methods is described.

The mathematical means to tackle problems such as those in ion–solid interactions were introduced in the last century, in the context of kinetic theory. This theory allows the determination of macroscopic properties of matter from a knowledge of the elementary atomic interactions. One of the most outstanding results of this theory is the Boltzmann transport equation and we will discuss in this chapter the derivation of the equation and how it may be used to solve a variety of problems concerning the penetration of ions in solids.

5.1.2 *The forward transport equation*

In this section the Boltzmann transport equation in the so-called forward form is derived. To this end we assume that ion bombardment gives birth to a distribution of particles,

$$F(r, v, t) \tag{5.1}$$

whose position vectors are r and velocities v at time t.

This function represents the number of particles per unit volume *resulting* from the interaction of the ions with the target so the distribution of particles present *before* bombardment is excluded. It is assumed that F is small compared with the atomic density, so that the probability for a target atom to be hit more than once is negligibly small.

We assume that the motion of the particles can be approximated by small (straight-line) displacements, within which only *one* scattering event may occur. Scattering will be regarded as an instantaneous process and it will be the only mechanism capable of changing the particle velocity, i.e., no force-field of any kind will be present in our derivations. We also assume that the binary collision approximation (BCA) holds and that the collision partner is always a stationary target atom. Finally, we introduce a function $S(r, v, t)$ representing the distribution of ions arriving per second into the target and $P(v \rightarrow v')$, which gives the probability per unit time of changing from velocity v to v'.

The function $F(r, v, t)$ satisfies the following probability balance equation:

$$F(r, v, t+\delta t) = F(r - \delta t\, v, v, t) \left(1 - \delta t \sum_{v'} P(v \rightarrow v') \right)$$
$$+ \sum_{v'} F(r - \delta t\, \hat{v}, v', t)\, \delta t\, P(v' \rightarrow v) + \delta t\, S(r, v, t), \tag{5.2}$$

where $\hat{v}$ is an intermediate velocity, between v and v'.

Particles may arrive at the point (r, v) in the phase space by one of the following paths.

(i) From particles within $(r - \delta t\, v, v)$ which do not scatter during δt. This is expressed by the first term on the right-hand side of equation (5.2), where

$$1 - \delta t \sum_{v'} P(v \rightarrow v')$$

denotes the probability of having no scattering within δt.

(ii) From particles at $(r - \delta t\, \hat{v}, v')$, which have velocity v after scattering,

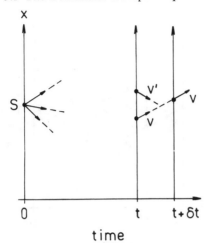

Fig. 5.1 The one-dimensional approximation to the events taking place within δt according to the forward picture in the transport equation.

as depicted by the second term on the right-hand side of equation (5.2).

(iii) From a source that, at site r, has emitted particles with velocity v, as denoted by the last term on the right-hand side of (5.2).

We illustrate in Figure 5.1 the probability balance in equation (5.2) for a one-dimensional case. There we can see particles arriving at (x, v) at $t + \delta t$ from different sites in the phase space following one of the three paths.

The transition probability is given in terms of the scattering cross-section, σ, by,

$$P(v \rightarrow v') = N|v|\sigma(v, v'), \qquad (5.3)$$

N being the atomic density of the target. Replacing summations by integrals, equation (5.2) becomes

$$F(r, v, t + \delta t) = F(r - \delta t \, v, v, t)\left(1 - \delta t \, N|v|\int dv' \, \sigma(v, v')\right)$$

$$+ \delta t \, N \int |v'|\sigma(v', v)\, dv' \, F(r - \delta t \hat{v}, v', t) + \delta t \, S(r, v, t). \qquad (5.4)$$

In the limit as $\delta t \rightarrow 0$ this gives the forward transport equation

$$\frac{\partial F(r, v, t)}{\partial t} = -(v \cdot \nabla_r)F(r, v, t) - N|v|F(r, v, t)\int dv' \, \sigma(v, v')$$

$$+ N \int |v'| \, dv' \, F(r, v', t)\sigma(v', v) + S(r, v, t). \qquad (5.5)$$

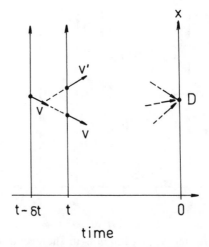

time

Fig. 5.2 The one-dimensional approximation to the events taking place within δt according to the backward picture in the transport equation.

5.1.3 The backward transport equation

In this section an equation similar to (5.5) is derived, in which the variables become those of the initial condition. Suppose that one has already obtained a solution of the forward equation for a particle, which at time t_0 resulted from a well-defined position and velocity, i.e., r_0 and v_0 or, more precisely,

$$S(r, v, t) = \delta(t - t_0)\delta(r - r_0)\delta(v - v_0),$$

where $\delta(x)$ is the Dirac delta function. We denote such a particular solution as

$$G(r, v, t | r_0, v_0, t_0),$$

where $t_0 < t$ because it is demanded by causality.

The initial distribution of particles $F^\dagger(r_0, v_0, t)$ is related to G by the following integral:

$$F^\dagger(r_0, v_0, t_0) = \int dr\, dv\, dt\, G(r, v, t | r_0, v_0, t_0)\, \mathscr{D}(r, v, t). \qquad (5.6)$$

The function $\mathscr{D}$ represents a detection procedure and mathematically expresses the position, sensitivity and time at which a detector becomes operative. Since the variables in this formulation are those in the initial state, time must necessarily go from future to past, i.e., *backwards*. For this reason, the transport equation derived using this approach is called the backward Boltzmann equation. We shall see later that there are computational advantages to this formulation.

Suppose that we know $F^\dagger(r_0, v_0, t_0)$ for a given time t_0. We can obtain $F^\dagger$ for $t_0 - \delta t_0$ ($\delta t_0 > 0$) following the same type of arguments as in the forward case. In the backward formulation the equation analogous to (5.2) is

$$F^\dagger(r_0, v_0, t_0 - \delta t_0) = F^\dagger(r_0 + \delta t_0\, v_0, t_0) \left(1 - \delta t \sum_{v_0'} P(v_0 \rightarrow v_0') \right)$$

$$+ \sum_{v_0'} F^\dagger(r_0 + \delta t_0\, \hat{v}_0, v_0', t_0)\delta t_0\, P(v_0 \rightarrow v_0') + \delta t\, \mathscr{D}(r_0, v_0, t_0), \tag{5.7}$$

where the terms on the right-hand side correspond to particles undergoing no scattering, scattering or contributing directly to the detector $\mathscr{D}$, respectively. Analogously to the forward equation, one can expand the function $F^\dagger$ in a power series in δt_0 and take the limit as $\delta t_0 \rightarrow 0$ to obtain

$$-\frac{\partial F^\dagger(r, v, t)}{\partial t} = (v \cdot \Delta_r)F^\dagger(r, v, t) - N|v|F^\dagger(r, v, t) \int dv'\, \sigma(v, v')$$

$$+ N|v| \int dv'\, F^\dagger(r, v', t)\sigma(v, v') + \mathscr{D}(r, v, t), \tag{5.8}$$

where we have now omitted the zero subscript used for the initial variables.

One of the advantages of the backward form is that the integrals on the right-hand side of (5.8) can be grouped together in the form

$$N|v| \int dv'\, \sigma(v, v') \left[F^\dagger(r, v', t) - F^\dagger(r, v, t) \right].$$

This grouping can often simplify the mathematics used in the solution, but it must be pointed out that the function $F^\dagger$ is not necessarily a probability distribution.

5.1.4 The time-independent Boltzmann equation

In most experiments we observe the accumulated effects of ion bombardment and so a full, time-dependent solution of the Boltzmann equation (BE) is often not required. In order to eliminate time one can assume that the beam (source) or detectors become active only during a certain period of time. Hence, the accumulated values can be obtained by integrating F or $F^\dagger$ over time. The corresponding time-independent transport equations are obtained by integrating equations (5.5) and (5.8) respectively.

We can thus introduce

$$F(\mathbf{r}, \mathbf{v}) = \int_{-\infty}^{+\infty} dt\, F(\mathbf{r}, \mathbf{v}, t),$$

$$F^{\dagger}(\mathbf{r}, \mathbf{v}) = \int_{-\infty}^{+\infty} dt\, F^{\dagger}(\mathbf{r}, \mathbf{v}, t).$$

Integrating (5.5) and (5.8), one obtains

(1) the time-independent forward transport equation

$$(\mathbf{v} \cdot \nabla_{\mathbf{r}}) F(\mathbf{r}, \mathbf{v}) = -F(\mathbf{r}, \mathbf{v}) N |\mathbf{v}| \int d\mathbf{v}'\, \sigma(\mathbf{v}, \mathbf{v}')$$
$$+ N \int d\mathbf{v}' |\mathbf{v}'| \sigma(\mathbf{v}', \mathbf{v}) F(\mathbf{r}, \mathbf{v}') + \mathcal{S}(\mathbf{r}, \mathbf{v}), \tag{5.9}$$

(2) the time-independent backward transport equation

$$(\mathbf{v} \cdot \nabla_{\mathbf{r}}) F^{\dagger}(\mathbf{r}, \mathbf{v}) = -N |\mathbf{v}| F^{\dagger}(\mathbf{r}, \mathbf{v}) \int d\mathbf{v}' \sigma(\mathbf{v}, \mathbf{v}')$$
$$+ N |\mathbf{v}| \int d\mathbf{v}'\, \sigma(\mathbf{v}, \mathbf{v}') F^{\dagger}(\mathbf{r}, \mathbf{v}') + \mathcal{D}(\mathbf{r}, \mathbf{v}). \tag{5.10}$$

Here,

$$\mathcal{S}(\mathbf{r}, \mathbf{v}) = \int_{-\infty}^{+\infty} dt\, \mathcal{S}(\mathbf{r}, \mathbf{v}, t),$$

$$\mathcal{D}(\mathbf{r}, \mathbf{v}) = \int_{-\infty}^{+\infty} dt\, \mathcal{D}(\mathbf{r}, \mathbf{v}, t).$$

These two forms of the BE will allow us to address problems of the type described at the beginning of this chapter.

5.1.5 Remarks

The assumptions used in previous derivations lead us to the so-called linear approximation to the Boltzmann equation. They imply that the interaction between the incoming particles and their own perturbations in the target are explicitly excluded. The linear approximation is valid only in the early stage of the ion–solid interaction. It cannot describe non-binary collisions, penetration in crystals or the latter stages of a cascade, during which collisions between moving particles may be important. Despite these limitations, the linear BE can predict a number of important experimental observables and has the advantage of being more amenable to analytical treatment than its full non-linear counterpart. Which of the two equivalent forms, the forward form (FBE) or the backward form (BBE), to use depends on the particular problem under investigation. The reader interested in details about derivation of the BE is referred to more specific literature (Williams, 1971, Gardiner, 1985, Case and Zweifel, 1967).

5.2 Ion penetration

5.2.1 The energy loss distribution

In this section we shall study the passage of a mono-energetic and well-collimated beam of ions through a thin solid film. The energy of the bombarding ions and the thickness of the film are chosen in such a way that particles pass through with relatively small energy losses and negligibly small deflections. The energy spectra of the transmitted ions can thus be measured and various parameters characterizing such spectra can be readily obtained. The quantity we would like to determine is a function $\phi(z, E)$ representing the number of particles traversing a foil of thickness z, with an energy E. Because z and E refer to the *final* state of the ions, it appears that the forward transport equation should be used.

The flux of ions through a plane perpendicular to the velocity at a certain time t with position r and velocity v can be obtained as $|v|F(r, v, t)$. After integrating over time we have $|v|F(r, v)$ for the integrated flux. If one assumes that, at the origin of coordinates, ϕ_0 ions have arrived with velocity $v_0 k$ then the source term becomes

$$S(r, v) = \phi_0 \delta(r)\delta(v - v_0 k),$$

k being the unit vector along the z direction.

In penetration experiments, films can be chosen thin enough that the energy loss is small compared with the bombarding energy. We can thus assume that the velocity along the beam direction changes by only a small amount during penetration, i.e. $v_z \simeq v_0$, v_0 being the initial ion velocity. Analogously, one can assume that $|v_x|, |v_y| \ll v_0$; hence, $|v| \simeq v_0$.

The transport equation for such a function can be readily obtained from equation (5.9), replacing $|v|F(r, v)$ by $\phi(r, v)$, giving

$$\frac{1}{v_0}(v \cdot \nabla_r)\phi(r, v) = -\phi(r, v)N \int dv' \, \sigma(v, v')$$

$$+ N \int dv' \, \sigma(v', v)\phi(r, v') + \phi_0 \delta(r)\delta(v - v_0 k). \tag{5.11}$$

The scattering cross-section in the first integral refers to the velocity *before* the collision, whereas the second one refers to the *final* velocity. Therefore,

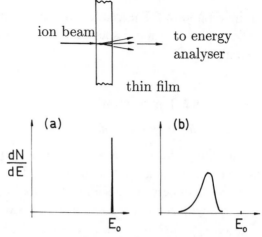

Fig. 5.3 Penetration of a mono-energetic and well-collimated ion beam in a thin solid film. (a) Energy spectrum of ions *before* penetration. (b) Energy spectrum of ions *after* traversing the foil.

the first integral on the right-hand side of (5.11) accounts for ions that are lost from the flux because of transitions to smaller velocities whereas the last integral represents ions with velocities greater than v resulting in a velocity v after scattering.

For a one-dimensional distribution we can integrate our lateral position x and y to give

$$\frac{v_z}{v_0}\frac{\partial \phi(z,v)}{\partial z} = -\phi(z,v)N\int dv'\,\sigma(v,v')$$
$$+ N\int dv'\,\sigma(v',v)\phi(z,v') + \phi_0\delta(z)\delta(v-v_0k), \tag{5.12}$$

where

$$\phi(z,v) = \int dx\,dy\,\phi(r,v).$$

As $|v| \sim v_0$ we can actually write

$$\frac{\partial \phi(z,v)}{\partial z} = -N\phi(z,v)\int dv'\,\sigma(v,v')$$
$$+ N\int dv'\,\sigma(v',v)\phi(z,v') + \phi_0\delta(z)\delta(v-v_0k). \tag{5.13}$$

Now, using energy in place of velocity we can define

$$\phi(z,v)\,dv \rightarrow \phi(z,E,n)\,dE\,dn, \tag{5.14}$$

$$\sigma(\boldsymbol{v}, \boldsymbol{v}') \, \mathrm{d}\boldsymbol{v}' \rightarrow \sigma(E, T) f(\boldsymbol{n}, \boldsymbol{n}') \, \mathrm{d}T \, \mathrm{d}\boldsymbol{n}', \tag{5.15}$$

$$\delta(z)\delta(\boldsymbol{v} - v_0\boldsymbol{k}) \rightarrow \delta(z)\delta(E - E_0)\delta(\boldsymbol{n} - \boldsymbol{k}), \tag{5.16}$$

where f is a function relating the direction of motion *before* and *after* scattering, viz. $\boldsymbol{n}$ and $\boldsymbol{n}'$, respectively. Analogously, $\sigma(E, T)$ is the scattering cross-section as a function of incoming energy E and energy loss T. The function f is

$$f(\boldsymbol{n}, \boldsymbol{n}') = \frac{\delta(\boldsymbol{n} \cdot \boldsymbol{n}' - \cos \psi)}{2\pi}, \tag{5.17}$$

where $\cos \psi$ is the scattering angle in the laboratory frame of reference. If, as before, we assume that $\cos \psi \simeq 1$, then it follows from equation (5.17) that $\boldsymbol{n} \cdot \boldsymbol{n}' = 1$, so scattering does not change the direction of motion. For normal incidence therefore

$$\phi(z, E, \boldsymbol{n}) = \phi(z, E)\delta(\boldsymbol{n} - \boldsymbol{k}).$$

Substituting into equation (5.13) and integrating over $\boldsymbol{n}$ gives

$$\begin{aligned} \frac{\partial \phi(z, E)}{\partial z} &= -\phi(z, E)N \int \mathrm{d}T \, \sigma(E, T) \\ &+ N \int \mathrm{d}T \, \sigma(E + T, T)\phi(z, E + T) + \phi_0\delta(z)\delta(E - E_0), \end{aligned} \tag{5.18}$$

where E_0 is the bombarding energy.

Making a further assumption that the scattering cross-section $\sigma(E, T)$ is strongly peaked at small energy loss, we can replace $\sigma(E + T, T)$ by $\sigma(E, T)$ to give

$$\frac{\partial \phi(z, E)}{\partial z} = N \int \mathrm{d}T \, \sigma(E, T)[\phi(z, E + T) - \phi(z, E)] + \phi_0\delta(z)\delta(E - E_0). \tag{5.19}$$

The same argument also applies to the function $\phi(z, E + T)$ which can therefore be expanded as

$$\phi(z, E + T) = \phi(z, E) + T\frac{\partial \phi(z, E)}{\partial E} + \frac{T^2}{2}\frac{\partial^2 \phi(z, E)}{\partial E^2} + \dots. \tag{5.20}$$

Introducing such an expansion into equation (5.19) one finally obtains

$$\frac{\partial \phi(z, E)}{\partial z} = NS(E)\frac{\partial \phi(z, E)}{\partial E} + \frac{N\Omega^2(E)}{2}\frac{\partial^2 \phi(z, E)}{\partial E^2} + \dots, \tag{5.21}$$

with

$$\phi(z = 0, E) = \phi_0\delta(E - E_0), \tag{5.22}$$

where $S(E)$ and $\Omega^2(E)$ are the stopping and energy straggling cross-sections, respectively.

$$S(E) = \int dT \; T \, \sigma(E, T), \tag{5.23}$$

$$\Omega^2(E) = \int dT \, \sigma(E, T) T^2. \tag{5.24}$$

The cross-section entering equations (5.23) and (5.24) may include both elastic and inelastic scattering. Assuming that they are two independent processes

$$\sigma(E, T) = \sigma_n(E, T) + \sigma_e(E, T).$$

Hence,

$$S(E) = S_e(E) + S_n(E),$$

as well as

$$\Omega^2(E) = \Omega_e^2(E) + \Omega_n^2(E)$$

and so forth. Here the subscripts e and n stand for electronic (or inelastic) and nuclear (or elastic), respectively.

Higher moments of the energy loss cross-section are required if more terms in the expansion (5.20) are included. The first and second moments shown above are, however, the most relevant in characterizing the interaction of ions with solids.

Following our assumption that the energy of the ions E changes only within a small fraction of the initial energy E_0, we can therefore evaluate the stopping and straggling at E_0 by solving

$$\frac{\partial \phi(z, E)}{\partial z} = N S(E_0) \frac{\partial \phi(z, E)}{\partial E} + \frac{N \Omega^2(E_0)}{2} \frac{\partial^2 \phi(z, E)}{\partial E^2}, \tag{5.25}$$

where terms greater than second order have been ignored. The solution with the boundary condition $\phi(0, E_0) = \phi_0 \delta(E - E_0)$ is

$$\phi(z, E) = \frac{\phi_0}{[2\pi z N \Omega(E_0)]^{\frac{1}{2}}} \exp\left(-\frac{[E_0 - zNS(E_0) - E]^2}{2zN\Omega(E_0)}\right), \tag{5.26}$$

or, in terms of the energy loss, $\Delta E = E_0 - E$,

$$\phi(z, \Delta E) = \frac{\phi_0}{[2\pi z N \Omega(E_0)]^{\frac{1}{2}}} \exp\left(-\frac{[\Delta E - zNS(E_0)]^2}{2zN\Omega(E_0)}\right). \tag{5.27}$$

This Gaussian approximation is expected to describe reasonably well the spectra of ions transmitted through thin films. However, one should be cautious about using this result in cases in which the energy loss becomes comparable with, say, 20% of the initial energy.

Equations (5.26) and (5.27) were obtained assuming small energy losses in a scattering event. This approximation is known as the diffusive approximation. Large energy losses can occur in thicker films.

Introducing the moments of the energy loss spectra,

$$\langle E^n \rangle(z) = \int dE \, E^n \phi(z, E),$$

one can easily see that by multiplying equation (5.25) by E^n and integrating over energy E, a set of coupled equations can be obtained for the moments,

$$\frac{\partial \langle E^n \rangle(z)}{\partial z} = -nNS(E_0)\langle E^{n-1} \rangle(z)$$
$$+ n(n-1)\frac{N\Omega^2(E_0)}{2}\langle E^{n-2} \rangle(z) + \ldots, \tag{5.28}$$

with

$$\langle E^n \rangle(z = 0) = E_0^n.$$

It is easily verified that

$$\langle E^0 \rangle(z) = 1,$$
$$\langle E^1 \rangle(z) = E_0 - zNS(E_0), \tag{5.29}$$
$$\langle E^2 \rangle(z) = \langle E^1 \rangle^2(z) + zN\Omega^2(E_0)$$

and so forth. These formulae can be inverted to give the stopping and straggling cross-sections:

$$S(E_0) = \frac{\langle E^0 \rangle - \langle E^1 \rangle}{zN},$$
$$\Omega^2(E_0) = \frac{\langle E^2 \rangle - \langle E^1 \rangle^2}{zN}. \tag{5.30}$$

These formulae can then be used in conjunction with energy loss measurements through thin films to determine S and Ω^2.

A more accurate expression for ϕ can be obtained by including higher moments of the energy loss into calculations. In this way one can reconstruct the energy loss distribution beyond the Gaussian approximation, by including skewness and kurtosis.

5.2.2 Range distribution: small deflections

In many ion-beam applications the mean depth at which the bombarding particles come to rest is required. The distribution of stopped ions as a

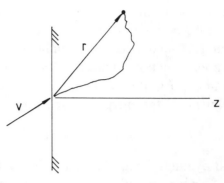

Fig. 5.4 An ion with velocity v at the origin may become stopped within $(r, \mathrm{d}r)$.

function of depth is called the range distribution, which is an especially useful quantity to know in doping semiconductors by ion-implantation, for which concentration as a function of depth should be predictable in order to choose the appropriate bombarding conditions.

We introduce a function $F(v, r)$ denoting the probability that a particle at the origin moving with velocity v may become stopped within position $(r, \mathrm{d}r)$. Note that this function is not the same as the solution of the FBE introduced earlier.

The first difficulty here arises from the mixed nature of the variables in the function $F(v, r)$, v is part of the *initial* state of the particle, r stands for the *final* position. None of the transport equations derived in Section 5.1, therefore, apply to this case.

Such a difficulty can be avoided if one assumes a *translation symmetry*. If properties of the target do not depend on position, as for example in an infinite and homogeneous target, then, such a stringent distinction between final and initial position disappears. In this case one can refer the position of the incoming ion from the point where we would like to have it stopped and pass from *final* to *initial* variables by a simple change of sign, i.e., $r \rightarrow -r$.

Thus the backward form of the BE can be used by writing

$$F(v, r) = F^{\dagger}(-r, v). \tag{5.31}$$

Note that previous assumptions converted the target into an infinite medium, for which there is no longer a surface. This is not a serious limitation since in most cases ions penetrate well below the surface and only few are affected by the presence of the vacuum.

According to equation (5.31) one can substitute $F^{\dagger}$ by F in equation (5.10)

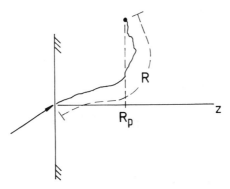

Fig. 5.5 The difference between range R and projected range R_p.

then, after some re-arrangements, one obtains

$$(\boldsymbol{v} \cdot \nabla_{\boldsymbol{r}}) F(\boldsymbol{v}, \mathbf{r}) = N|\boldsymbol{v}| \int \mathrm{d}\boldsymbol{v}' \, \sigma(\boldsymbol{v}, \boldsymbol{v}') [F(\boldsymbol{v}', \mathbf{r}) - F(\boldsymbol{v}, \mathbf{r})]$$
$$+ \mathscr{D}(-\mathbf{r}, \boldsymbol{v}). \tag{5.32}$$

Since the detector must be able to localise stopped ions around the origin and cannot depend on the direction of motion of ions, one can assume that

$$\mathscr{D}(-\mathbf{r}, \boldsymbol{v}) = \delta(\mathbf{r})\mathscr{D}(|\boldsymbol{v}|). \tag{5.33}$$

Furthermore, if one only needs the distribution of stopped particles with depth, the function $F(\boldsymbol{v}, \mathbf{r})$ can be integrated over x and y,

$$F(\boldsymbol{v}, z) = \int \mathrm{d}x \, \mathrm{d}y \, F(\boldsymbol{v}, \mathbf{r}). \tag{5.34}$$

Analogously,

$$v_z \frac{\partial F(\boldsymbol{v}, z)}{\partial z} = N|\boldsymbol{v}| \int \mathrm{d}\boldsymbol{v}' \, \sigma(\boldsymbol{v}, \boldsymbol{v}') [F(\boldsymbol{v}', z) - F(\boldsymbol{v}, z)] + \delta(z)\mathscr{D}(|\boldsymbol{v}|). \tag{5.35}$$

The range along a direction perpendicular to the surface is often known as *projected range*. It is different from *range*, which denotes the mean distance followed by the ion along its own path. Projected range is a more useful quantity than range since it relates to the depth at which particles come to rest. In the rest of the chapter the word range will imply projected range unless the contrary is stated.

   In order to use a more standard notation, c.f. Sigmund (1972), we shall write the range distribution in terms of energy E and direction cosine $\eta = \cos\theta = v_z/|\boldsymbol{v}|$, viz. $F(E, \eta, z)$. Similarly, the scattering cross-section will be written using energy and energy loss T, as in equations (5.14)–(5.16). The scattering cross-section entering above may include both elastic and inelastic interactions. For the time being, however, we need not be more specific.

With all these changes one finally obtains

$$\eta \frac{\partial F(E,\eta,z)}{\partial z} = \delta(z)\mathscr{D}(E)$$
$$+ N \int dT\, d\mathbf{n}'\, \sigma(E,T) f(\mathbf{n},\mathbf{n}')[F(E-T,\eta',z) - F(E,\eta,z)]. \tag{5.36}$$

Clearly, η and η' are the components of $\mathbf{n}$ and $\mathbf{n}'$ along the z axis, respectively. Here, η' depends on η, E and T according to equation (5.17). After integration over x and y, the function F cannot depend on any component of the initial ion direction other than $\cos\theta = \eta$.

Equation (5.36) still poses a rather difficult mathematical problem. It embodies an integro-differential equation involving three variables: energy, position and direction cosine.

Spatial moments. One can reduce equation (5.36) to a more amenable form by means of the so-called method of moments. The spatial moments are defined by

$$F^{(n)}(E,\eta) = \int dz\, z^n\, F(E,\eta,z). \tag{5.37}$$

Thus integrating (5.36) over z gives

$$-n\eta F^{(n-1)}(E,\eta) = \delta_{n,0} D(E)$$
$$+ N \int dT\, d\mathbf{n}'\, \sigma(E,T) f(\mathbf{n},\mathbf{n}')[F^{(n)}(E-T,\eta') - F^{(n)}(E,\eta)]. \tag{5.38}$$

Here $\delta_{n,0}$ is the Kronecker delta. Equation (5.38) is now an integral equation for $F^{(n)}$ in terms of $F^{(n-1)}$. Knowing $F^{(0)}$ enables the higher order moments to be derived recursively.

Using spatial moments introduces another additional advantage. Provided that the projected range has a single maximum and does not show a slowly decaying tail at large depth, the projected range distribution can be fairly well approximated by a Gaussian. In this case

$$F(E,\eta,z) \simeq \exp\left(-\frac{(z-\bar{z})^2}{2\sigma^2}\right),$$

where the mean depth $\bar{z}$ and the standard deviation, σ, or depth straggling, can be obtained from the first and second spatial moments as

$$\langle z \rangle = \frac{F^{(1)}(E, \eta)}{F^{(0)}(E, \eta)}, \tag{5.39}$$

$$\sigma^2 = \langle \Delta z^2 \rangle = \langle z^2 \rangle - \langle z \rangle^2, \tag{5.40}$$

with

$$\langle z^2 \rangle = \frac{F^{(2)}(E, \eta)}{F^{(0)}(E, \eta)}. \tag{5.41}$$

Deviations from Gaussian, if they are not large, can be corrected by including derivatives of the Gaussian function. One procedure known as the Edgeworth series (Cramér, 1958) demands the calculation of moments higher than the second. In fact, according to Edgeworth, any distribution, i.e. $F(z)$, can be written as

$$F(\xi) = \varphi(\xi) - \frac{1}{3!} \gamma_1 \varphi^{(3)} + \frac{1}{4!} \gamma_2 \varphi^{(4)}(\xi) + \dots,$$

where φ is the Gaussian function, $\varphi^{(n)}(\xi) = d^n \varphi(\xi)/d\xi^n$, with $\xi = (z - \langle z \rangle)/\sigma$, and

$$\gamma_1 = \frac{\mu_3}{\sigma^3},$$

$$\gamma_2 = \frac{\mu_4}{\sigma^4} - 3$$

with $\mu_n = \langle (z - \langle z \rangle)^n \rangle$, i.e. the centred moment of order n.

The coefficients γ_1 and γ_2 are also known as the skewness and kurtosis, respectively. They account for the asymmetry and the 'slenderness' of the distribution. A single maximum distribution that decays slower on the right- than on the left-hand side will have a positive skewness coefficient, whereas a distribution with a broad maximum but rapidly decaying on both sides will have a positive kurtosis. Note that this is a different definition of kurtosis than that used in the next chapter.

In Figure 5.6 we plot the range distribution of 120 keV Bi ions implanted into an Si target. The data follow fairly well a Gaussian shape. The errors can be reduced by using an Edgeworth expansion up to the third derivative. The result shown as a dashed line corresponds to an Edgeworth expansion with the skewness coefficient $\gamma_1 = 0.28$ obtained from experimental data. Chapter 6 examines this problem in more detail using Pearson and Johnson distributions.

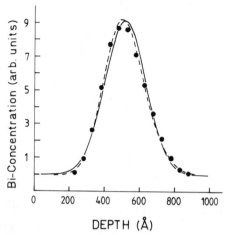

Fig. 5.6 Approximating the range distribution of 120 keV Bi on Si by a Gaussian function (continuous line). By using the Edgeworth series up to the third derivative term (dashed line), the fitting is improved.

Normalization. In order to proceed further with equation (5.38) we have two possibilities: either we find an expression for the detector term or the zeroth order spatial moment, $F^{(0)}(E, \eta)$, must be given. Given that ions must stop somewhere in the target, integration of $F(E, \eta, z)$ along all depths adds up to the certainty event, so

$$\int_{-\infty}^{+\infty} dz \, F(E, \eta, z) = F^{(0)}(E, \eta) = 1.$$

Heavy ions. The mathematical analysis can be simplified in the case of heavy ions for which scattering with target atoms seldom leads to large deflections, so $\cos \phi \simeq 1$. The directional cosine remains practically unchanged during penetration, thus $\eta \simeq 1$. In this case equation (5.37) can be written as

$$-nF^{(n-1)}(E) = N \int dT \, \sigma(E, T) \left[F^{(n)}(E - T) - F^{(n)}(E) \right] ; \; n > 0, \quad (5.42)$$

where we have replaced $F^{(n)}(E, \eta)$ by $F^{(n)}(E)$ since η is no longer a variable.

By introducing the power scattering cross-section (equation (2.70))

$$\sigma(E, T) = \frac{C_m}{E^m T^{1+m}},$$

the calculations can be simplified since solutions for the moments exist in the form

$$F^{(n)}(E) = A^{(n)} R_0^n$$

with $R_0 = E^{2m}/(N C_m)$, where $A^{(n)}$ is a dimensionless coefficient that can be readily obtained from equation (5.42) as

$$nA^{(n-1)} = A^{(n)} \int_0^\gamma \frac{dx}{x^{1+m}}[1 - (1-x)^{2nm}], \qquad (5.43)$$

where $\gamma = 4m_1 m_2/(m_1 + m_2)^2$, m_1 and m_2 being the ion and target masses, respectively. All the moments can be obtained recursively starting from $A^{(0)}$, which according to normalization satisfies

$$A^{(0)} = 1.$$

Hence, carrying out the integration one obtains

$$A^{(n)} = \frac{nA^{(n-1)}}{B_\gamma(1, -m) - B_\gamma(1 + 2nm, -m)},$$

$B_\gamma(x, y)$ being the incomplete gamma function (Abramowitz and Stegun, 1970),

$$B_\gamma(x, y) = \int_0^\gamma du \, u^{x-1}(1 - u)^{y-1}.$$

One can readily obtain the first spatial moments of the range distribution, namely, the mean range and range straggling, as

$$\langle z \rangle = A^{(1)} R_0, \qquad (5.44)$$

$$\langle (\Delta z)^2 \rangle = [(A^{(2)} - A^{(1)})^2] R_0^2. \qquad (5.45)$$

Reconstruction of the spatial distribution from the spatial moments, though possible, can only be done in an approximate fashion from a finite number of them. The interested reader can refer to more specific sources such as Shohat and Tamarkin (1943). Here, we shall limit ourselves to calculating some quantities characterizing the spatial distribution without reconstructing the exact distribution.

Given that in solving equation (5.38) we have neglected deflections, the previous results can be identified with those of range R, so

$$\langle R \rangle / R_0 = A^{(1)},$$
$$\langle (\Delta R)^2 \rangle / R_0^2 = A^{(2)} - (A^{(1)})^2 \qquad (5.46)$$

for the mean range and the range straggling, respectively.

5.2.3 *Range distribution: the general case*

In this section we shall derive general expressions for the distribution of stopped particles. We resume our analysis from equation (5.36), although deflections will not be neglected this time.

The mathematical difficulty posed by the presence of two coupled variables, i.e. energy and angle, can be efficiently handled by expansion in Legendre polynomials,

$$F(z, E, \eta) = \sum_{l=0}^{\infty} (2l + 1) F_l(z, E) P_l(\eta), \tag{5.47}$$

where

$$F_l(z, E) = \frac{1}{2} \int_{-1}^{+1} \mathrm{d}\eta \, F(z, E, \eta) P_l(\eta) \tag{5.48}$$

and P_l is the Legendre polynomial of degree l.

One can replace $F(z, E, \eta)$ by its Legendre expansion, (5.47), in equation (5.36) and use the following properties of the Legendre polynomials (Abramowitz and Stegun, 1970) to simplify.

(1) The recurrence theorem,

$$(l + 1) P_{l+1}(x) = (2l + 1) x P_l(x) - l P_{l-1}(x). \tag{5.49}$$

(2) The addition theorem,

$$P_l(\cos \phi) = P_l(\cos \theta) P_l(\cos \theta')$$
$$+ 2 \sum_{m=1}^{l} \frac{(l - m)!}{(l + m)!} P_l^m(\cos \theta) P_l(\cos \theta') \cos m\varphi,$$

where ϕ stands for the angle between two unit vectors with direction cosines $\cos \theta$ and $\cos \theta'$ and a relative azimuth φ, P_l^m being the associated Legendre function.

Accordingly, we can write equation (5.36) as

$$-\frac{\partial}{\partial z} [l F_{l-1}(z, E) + (l + 1) F_{l+1}(z, E)]$$
$$= (2l + 1) N \int \mathrm{d}T \, \sigma(E, T) [F_l(z, E) - F_l(z, E - T) P_l(\cos \psi)] \tag{5.50}$$

and introducing spatial moments as before gives

$$n[l F_{l-1}^{(n-1)}(E) + (l + 1) F_{l+1}^{(n-1)}(E)]$$
$$= (2l + 1) N \int \mathrm{d}T \, \sigma(E, T) [F_l^{(n)}(E) - F_l^{(n)}(E - T) P_l(\cos \psi)], \tag{5.51}$$

where ψ is the scattering angle in the laboratory frame of reference and, owing to normalization, one has

$$F_l^{(0)} = \delta_{l,0}. \tag{5.52}$$

As a first attempt to solve equation (5.51) one can neglect inelastic energy loss and use, again, the power cross-section approximation for the elastic scattering,

$$n[lF_{l-1}^{(n-1)}(E) + (l+1)F_{l+1}^{(n-1)}(E)]$$
$$= (2l+1)E^{-m}NC_m \int_0^{\gamma E} \frac{dT}{T^{1+m}} [F_l^{(n)}(E) - F_l^{(n)}(E-T)P_l(\cos\psi)]. \tag{5.53}$$

Similarly to equation (5.42) a solution exists in the form

$$F_l^{(n)}(E) = A_l^{(n)} R_0^n$$

whence the following recursive relation for the A terms can be obtained:

$$n[lA_{l-1}^{(n-1)} + (l+1)A_{l+1}^{(n-1)}] = (2l+1)A_l^{(n)}I_l^{(n)}(\gamma), \tag{5.54}$$

where $\gamma = 4m_1m_2/(m_1+m_2)^2$ as before. According to (5.52),

$$A_l^{(0)} = \delta_{l,0}, \tag{5.55}$$

$$I_l^{(n)}(\gamma) = \int_0^\gamma dx \, [1 - (1-x)^{2nm} P_l(\cos\psi)]x^{-1-m}. \tag{5.56}$$

Following equations (5.53)–(5.54), the mean depth and depth straggling are given by

$$\langle z \rangle / R_0 = A_1^{(1)}(\gamma)P_1(\eta), \tag{5.57}$$

$$\langle \Delta z^2 \rangle / R_0^2 = A_0^{(2)} + 5A_2^{(2)}P_2(\eta) - 3(A_1^{(1)})^2 P_1^2(\eta). \tag{5.58}$$

Following this procedure one can obtain higher moments, which would be required if the spatial distribution $F(z, E, \eta)$ is to be reconstructed beyond the Gaussian approximation. For example, for heavy ions and high bombarding energies the range profiles show strong deviations from a Gaussian shape, so that calculation of skewness and kurtosis coefficients would become necessary.

Despite the similarities between equations (5.56)–(5.58) and those obtained assuming no deflection in equations (5.44) and (5.45) (i.e. in both cases the distributions can be written using the same length scale R_0), in the general

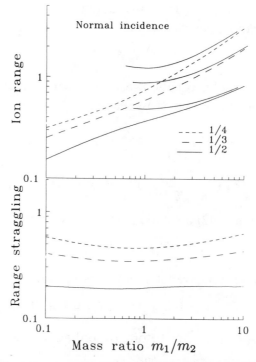

Fig. 5.7 Range and range straggling for normal incidence as functions of the mass ratio, m_1/m_2. Here m_1 and m_2 are the ion and target masses. Calculations were carried out using $m = 1/2$, $1/3$ and $1/4$ power potentials, respectively. For comparison, the corresponding mean ranges along the path appear as the thin continuous lines shown at large mass ratios.

case both the range and range straggling (standard deviation of the range) explicitly depend on the angle of incidence.

Figure 5.7 illustrates the scaled range and range straggling as functions of the mass ratio and various power laws for normal incidence. Observe that, while range becomes sensitive both to the mass ratio and to the power m, the range straggling is relatively insensitive to the mass ratio. The mean range along the path, i.e. equation (5.44), appears to approximate calculations including deflections fairly well.

Lateral spread. As pointed out by Winterbon, Sigmund and Sanders (1970), by using the results in the general case one can obtain the lateral distribution of the range distribution. In fact, since $\langle \Delta z^2 \rangle$ corresponds to $\langle \rho^2 \rangle = \langle x^2 \rangle + \langle y^2 \rangle$ in the case $\eta = 0$, one thus has

$$\langle \rho^2 \rangle / R_0^2 = A_0^{(2)}(\gamma) + 5 A_2^{(2)}(\gamma) P_2(0).$$

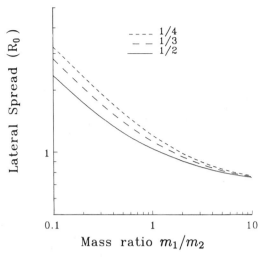

Fig. 5.8 Lateral spread in units of R_0 as a function of the mass ratio, m_1/m_2, and $m = 1/2, 1/3$ and $1/4$ power potentials, respectively.

Figure 5.8 shows the lateral spread of the range distribution in units of R_0. The lateral spread decreases with increasing ion mass. This is in agreement with the intuitive 'picture' that heavy ions will undergo small deviations. This lateral spread has important consequences when doping narrow p- or n-channels during fabrication of integrated circuits. In fact, the minimum lateral dimension of the semiconductor devices will ultimately be limited by the lateral spread of the dopant ion.

In order to be able to compare these expressions with experiment it is necessary to know the best value of the power law potential exponent m and the constant C_m.

Lindhard's scaling. Lindhard, Nielsen and Scharff, (1968) derived a *similarity law*, which allows one to find m in a very straightforward manner. This is possible because the scattering cross-section closely follows the 'universal' expression, defined in Chapter 2, equation (2.71), *et seq.*:

$$d\sigma = \frac{\pi a^2}{2t^{3/2}} f(t^{1/2}) \, dt^{1/2}.$$

Winterbon, Sigmund and Sanders (1970) found useful the following approximate expression:

$$f(t^{1/2}) = \lambda t^{1/2-m}[1 + (2\lambda t^{1-m})^q]^{-1/p}, \tag{5.59}$$

where λ, m and q are listed in Table 5.1 for Thomas–Fermi (TF) and Lenz–Jensen (LJ) interatomic potentials, respectively.

Table 5.1. *Parameters in the cross-section function* $f(t^{1/2})$

Potential	λ	m	q
Thomas–Fermi	1.309	0.333	0.667
Lenz–Jensen	2.92	0.191	0.512

If one also defines depth by

$$\rho = N\pi a^2 \gamma x, \tag{5.60}$$

the nuclear stopping power becomes

$$-\frac{d\epsilon}{d\rho}\bigg|_n = \frac{1}{\epsilon}\int_0^\epsilon f(x)\,dx = S_n(\epsilon). \tag{5.61}$$

Figure 5.9 depicts the nuclear stopping power calculated using expression (5.59) with parameters in Table 5.1. At large energies both TF and LJ yield nearly the same stopping power, whereas at lower energies they become quite different. This is because at higher energies the exchange of energy occurs mainly at small separation at which LJ and TF are nearly identical, whereas the much shorter range of the LJ potential causes the stopping power to be smaller than that of the TF potential at lower energies.

Also in Figure 5.9 we have plotted the stopping which results from the approximation

$$S_n(\epsilon) \simeq \frac{1}{2}\frac{\lambda_m}{1-m}\epsilon^{1-2m}, \tag{5.62}$$

with $\lambda_{1/2} = 0.327$, $\lambda_{1/3} = 1.309$ and $\lambda_{1/4} = 1.25$. It can be seen that $m = 0.5$ fits reasonably well the nuclear stopping power within $0.05 \leq \epsilon \leq 2$. At lower energies, $m = 1/3$ fits the stopping power for the Thomas–Fermi potential, whereas $m = 1/4$ appears more adequate in the case of the Lenz–Jensen potential.

Now, if one compares the stopping power in (5.63) with that obtained using the power potential (2.95) i.e.

$$-\frac{dE}{dx}\bigg|_n = \frac{NC_m\gamma^{1-m}}{1-m}E^{1-2m} \tag{5.63}$$

the coefficient C_m results,

$$C_m = \frac{\pi}{2}\lambda_m a^2 \left(\frac{m_1}{m_2}\right)^m \left(\frac{2Z_1Z_2e^2}{4\pi\varepsilon_0 a}\right)^{2m}. \tag{5.64}$$

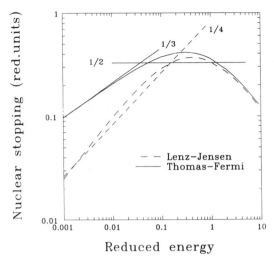

Fig. 5.9 Nuclear stopping power in Lindhard's reduced units (see text). Straight lines indicate power approximations.

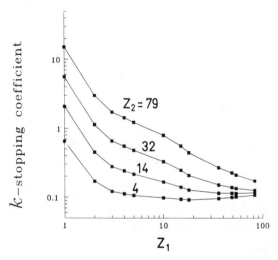

Fig. 5.10 Lindhard's electronic stopping coefficient as a function of the ion atomic number Z_1 and various atomic numbers in the target.

Here a is the screening length defined in Chapters 2 and 3 and thus R_0 and all moments in the range distribution can be completely determined.

Electronic energy loss. To take further advantage of Lindhard scaling, we can use the fact that the electronic stopping at low bombarding energies appears to be proportional to the ion velocity (see Chapter 4). This implies that, after translation to reduced units, the mean electronic energy loss per

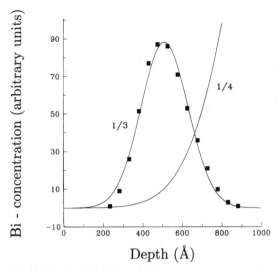

Fig. 5.11 Range profiles for 120 keV Bi implanted into amorphous Si for the power law potential with $m = 1/3$ and $m = 1/4$ from (5.57) and (5.58). Filled squares represent experimental results. In both cases inelastic energy loss was disregarded.

unit path can be written as

$$-\left.\frac{\mathrm{d}\epsilon}{\mathrm{d}\rho}\right|_e = k\epsilon^{1/2}, \tag{5.65}$$

where Lindhard's electronic stopping coefficient, k, is given by

$$k \simeq 0.0793 \frac{Z_1^{2/3} Z_2^{1/2} (A_1 + A_2)^{3/2}}{A_1^{3/2} A_2^{1/2} (Z_1^{2/3} + Z_2^{2/3})^{3/4}}.$$

We plot in Figure 5.10 the stopping coefficient, k, as a function of the ion atomic number, Z_1, for several targets. As one can see, k decreases with increasing atomic number of the ion and increases with increasing atomic number of the target.

In summary, having translated from E to ϵ one can not only find m but also, by the use of (5.65), one can see whether or not inelastic energy loss can be disregarded. For instance, returning to the example of 120 keV Bi into Si, the reduced energy becomes in this case $\epsilon = 8.3 \times 10^{-3}$. In addition, as $k \simeq 9.3 \times 10^{-3}$ it is clear that inelastic energy loss can be neglected as was the case for our calculations in the previous section.

In Figure 5.11 we plot the Bi range profile in Si together with the Gaussian approximation. It can readily be seen that $m = 1/3$ yields agreement with experiments, whereas $m = 1/4$ produces a range distribution that is clearly shifted towards large depths in the target.

5.2.4 High energy

As the energy of the ion increases the inelastic energy loss cannot be neglected. In this case one has to include the inelastic cross-section in equation (5.51). For most applications a satisfactory result can be obtained by using the following approximation:

$$\int dT\, \sigma_e(E, T)[F_l^{(n)}(E) - F_l^{(n)}(E - T)P_l(\cos\psi)] \approx S_e(E)\frac{dF_l^{(n)}(E)}{dE}, \quad (5.66)$$

where $S_e(E) = \int dT\, \sigma_e(E, T)T$ is the electronic stopping cross-section.

Introducing this term into equation (5.51) one arrives at the following set of integro-differential equations:

$$(2l+1)NS_e(E)\frac{dF_l^{(n)}(E)}{dE} = n[lF_{l-1}^{(n-1)}(E) + (l+1)F_{l+1}^{(n-1)}(E)]$$
$$- (2l+1)N\int_0^{\gamma E} dT\, \sigma_n(E, T)[F_l^{(n)}(E) - F_l^{(n)}(E - T)P_l(\cos\psi)]. \quad (5.67)$$

The initial conditions are the solutions of equation (5.53). The solution of (5.67) still requires a large computational effort. Such solutions are discussed in more detail in Chapter 6. In the next section an approximate method is derived, which considerably simplifies the solution procedure.

5.2.5 Numerical solution

If the differential nuclear scattering cross-section, $d\sigma_n(E, T)$, is peaked towards small T, then we can expand the term $F_l^{(n)}(E - T)$ by using a Taylor series. Performing this expansion, the integro-differential equation (5.67) is converted to the ordinary differential equation (ODE) given by

$$A_l(E)F_l^{(n)}(E) + B_l(E)\frac{dF_l^{(n)}(E)}{dE} - \sum_{k=2}^{\infty} B_l^k(E)\frac{(-1)^k}{k!}\frac{d^k F_l^{(n)}(E)}{dE^k} = S_l^n(E), \quad (5.68)$$

where

$$A_l(E) = N\int_0^{\gamma E}[1 - P_l(\cos\psi)]\, d\sigma_n(E, T),$$
$$B_l(E) = B_l^1(E) + NS_e(E),$$
$$B_l^k(E) = N\int_0^{\gamma E} T^k P_l(\cos\psi)\, d\sigma_n(E, T),$$

$$S_l^n(E) = n[lF_{l-1}^{(n-1)}(E) + (l+1)F_{l+1}^{(n-1)}(E)]/(2l+1). \quad (5.69)$$

For practical solutions of equation (5.69) the series in k must be finite. In order to compute moments up to second order it is not unreasonable to solve second-order ODEs, ie to truncate the series at $k = 2$. The three ODEs which are necessary in order to obtain second-order moments are, in operator notation,

$$D_1 F_1^{(1)}(E) = \frac{1}{3}, \ D_0 F_0^{(2)}(E) = 2F_1^{(1)}(E), \ D_2 F_2^{(2)}(E) = \frac{4}{5} F_1^{(1)}(E), \quad (5.70)$$

where

$$D_l = A_l(E) + B_l(E) \frac{d}{dE} - \frac{B_l^2(E)}{2} \frac{d^2}{dE^2}. \quad (5.71)$$

The equations (5.70) are coupled ODEs of the initial value type. The analysis of the characteristic equation associated with each of the equations (5.70) reveals a large positive eigenvalue and a small negative eigenvalue. Under these circumstances, the numerical solution of these equations cannot be approached directly by using well-known single- and multi-step techniques such as the Runge–Kutta and Adams methods.

However, the solution of such equations can be facilitated by recasting them into first-order ODEs designed for iterative solution. In this scheme, the true solution $F_l^{(n)}(E)$ is represented by an approximate solution, $F_{l,0}^{(n)}(E)$, plus a correction term, $F_{l,1}^{(n)}(E)$. An initial guess for $F_{l,0}^{(n)}(E)$ is obtained by solving the first-order ODE given by

$$A_l(E)F_{l,0}^{(n)}(E) + B_l(E) \frac{dF_{l,0}^{(n)}(E)}{dE} = S_{l,0}^n(E), \quad (5.72)$$

which is amenable to numerical solution by the forementioned techniques. Using the same numerical techniques, the correction term is obtained by solving another first-order ODE given by

$$A_l(E)F_{l,1}^{(n)}(E) + B_l(E) \frac{dF_{l,1}^{(n)}(E)}{dE} = S_{l,0}^n(E) - D_l F_{l,0}^{(n)}(E). \quad (5.73)$$

The true solution is then obtained by repeatedly updating the approximate solution, $F_{l,0}^{(n)}(E) := F_{l,0}^{(n)}(E) + F_{l,1}^{(n)}(E)$, while obtaining new correction terms by using equation (5.73). In practice, up to six iterations are necessary.

The computation of the integrals $A_l(E)$ and $B_l^k(E)$ can present difficulties on a small computer. Such integrals can be approximated by expanding the terms involving $\cos \psi$, where

$$\cos \psi = \left(1 - \frac{(1+A)}{2} \frac{T}{E}\right) \left(1 - \frac{T}{E}\right)^{-\frac{1}{2}}, \ A = \frac{m_2}{m_1}$$

$$\approx 1 - \frac{A}{2}(T/E) + \frac{1-2A}{8}(T/E)^2 + O((T/E)^3), \quad (5.74)$$

to yield a series of energy loss moments. If the energy loss moments up to second-order are retained, then the integrals are approximated by using

$$A_1(E) \approx N \left(\frac{AS_n(E)}{2E} - \frac{(1-2A)\Omega_n^2(E)}{8E^2} \right),$$

$$B_1(E) \approx N \left(S_t(E) - \frac{A\Omega_n^2(E)}{2E} \right),$$

$$A_0(E) = 0, \quad B_0(E) \approx NS_t(E),$$

$$A_2(E) \approx N \left(\frac{3AS_n(E)}{2E} - \frac{3(1-A)^2\Omega_n^2(E)}{8E^2} \right),$$

$$B_2(E) \approx N \left(S_t(E) - \frac{3A\Omega_n^2(E)}{2E} \right), \tag{5.75}$$

$$B_1^2(E) \approx B_0^2(E) \approx B_2^2(E) \approx N\Omega_n^2(E),$$

where

$$S_n(E) = \int_0^{\gamma E} T \, d\sigma_n(E, T), \quad S_t(E) = S_n(E) + S_e(E),$$

$$\Omega_n^2(E) = \int_0^{\gamma E} T^2 \, d\sigma_n(E, T). \tag{5.76}$$

Finally, the use of analytic fitting formulae for $S_n(E)$, $S_e(E)$ and $\Omega_n^2(E)$ alleviates the need for numerical integration. Fitting formulae for the ZBL interatomic potential are to be found in the literature (Ziegler, Biersack and Littmark, 1985).

It should be noted that, whereas the exact integral coefficient for $B_2(E)$ is always positive, the approximate coefficient, obtained when $m_1 \leq m_2$, passes through zero at certain energies. In such cases the numerical solution of equations (5.72) and (5.73) increases in complexity when using approximate coefficients. Specifically, when the equations are rearranged for numerical solution by isolating the first derivative on the left-hand side (i.e. the explicit form of the first-order ODE), $B_2(E)$ becomes the denominator of both source terms. Thus, any zeros in $B_2(E)$ manifest themselves as poles on the right-hand side. However, the ultimate accuracy of the true solution is independent of the approximations used for $A_l(E)$ and $B_l(E)$ on the left-hand side of equations (5.72) and (5.73). Simply by using the first-order energy loss moments for the coefficients on the left-hand side, the coefficient $B_2(E)$ is always positive and no singularities can occur on the right-hand side. The resulting equations, called the Kent optimised range algorithm (Bowyer,

Ashworth and Oven, 1994; Ashworth, Bowyer and Oven, 1995), for the range moments $F_1^{(1)}(E)$, $F_0^{(2)}(E)$ and $F_2^{(2)}(E)$ are (i) the initial guess,

$$
\begin{aligned}
\frac{dF_{1,0}^{(1)}(E)}{dE} &= \left(\frac{1}{3} - \frac{NAS_n(E)}{2E} F_{1,0}^{(1)}(E)\right)\frac{1}{NS_t(E)}, \\
\frac{dF_{0,0}^{(2)}(E)}{dE} &= \frac{2F_{1,0}^{(1)}(E)}{NS_t(E)}, \\
\frac{dF_{2,0}^{(2)}(E)}{dE} &= \left(\frac{4}{5}F_{1,0}^{(1)}(E) - \frac{3NAS_n(E)}{2E} F_{2,0}^{(2)}(E)\right)\frac{1}{NS_t(E)}
\end{aligned}
\tag{5.77}
$$

and (ii) the correction term

$$
\begin{aligned}
\frac{dF_{1,1}^{(1)}(E)}{dE} &= \left(\frac{1}{3} - D_1 F_{1,0}^{(1)}(E) - \frac{NAS_n(E)}{2E} F_{1,1}^{(1)}(E)\right)\frac{1}{NS_t(E)}, \\
\frac{dF_{0,1}^{(2)}(E)}{dE} &= \left[2F_{1,0}^{(1)}(E) - D_0 F_{0,0}^{(2)}(E)\right]\frac{1}{NS_t(E)}, \\
\frac{dF_{2,1}^{(2)}(E)}{dE} &= \left(\frac{4}{5}F_{1,0}^{(1)}(E) - D_2 F_{2,0}^{(2)}(E) - \frac{3NAS_n(E)}{2E} F_{2,1}^{(2)}(E)\right)\frac{1}{NS_t(E)}.
\end{aligned}
\tag{5.78}
$$

Finally, the projected range, $\langle z\rangle(E)$, the vertical standard deviation, $[\langle \Delta z^2\rangle(E)]^{1/2}$, and the lateral standard deviation, $[\langle \Delta x^2\rangle(E)]^{1/2}$, are obtained from the moments $F_l^{(n)}(E)$ by using (Winterbon, Sigmund and Sanders, 1970)

$$
\begin{aligned}
\langle z\rangle(E) &= 3F_1^{(1)}(E), \\
\langle \Delta z^2\rangle(E) &= F_0^{(2)}(E) + 5F_2^{(2)}(E) - [\langle z\rangle(E)]^2, \\
\langle \Delta x^2\rangle(E) &= F_0^{(2)}(E) - \frac{5}{2}F_2^{(2)}(E).
\end{aligned}
\tag{5.79}
$$

The initial conditions for the moments $F_1^{(n)}(E)$ can be obtained from equation (5.54) (i.e. by using power law potentials). The treatment of multi-component targets, by extension of equations (5.77) and (5.78) has been discussed elsewhere (Bowyer, Ashworth and Oven, 1994). A stand alone FORTRAN computer code, KORAL, that implements equations (5.77) and (5.78) is available[1]. The first three range moments obtained from KORAL, for As implanted into a-Si, are shown in Figure 5.12.

[1] From D. G. Ashworth, Electronic Engineering Laboratory, The University of Kent, Canterbury, Kent CT2 7NT, UK.

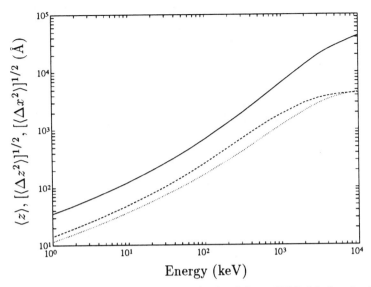

Energy (keV)

Fig. 5.12 The first three range moments obtained from KORAL for As implanted into a-Si: ——, $\langle z \rangle$; – – ––, and $[\langle \Delta z^2 \rangle]^{1/2}$; and $\cdots$, $[\langle \Delta x^2 \rangle]^{1/2}$.

The most widely used computer code for the solution of approximate transport equations is the Projected Range Algorithm, PRAL (Biersack, 1981). This code uses *simple* first-order ODEs to obtain second-order moments. This technique is known to be inaccurate (Furukawa, Matsumara and Ishiwara, 1972). Furthermore, it has been shown (Ashworth, Bowyer and Oven, 1991) that the detailed derivation of PRAL (Biersack, 1982) is incorrect. The relative errors of the codes KORAL and PRAL with respect to the Monte Carlo code TRIM85 are shown in Table 5.2. Also shown are the relative errors of the full transport equation solver KUBBIC (Bowyer, Ashworth and Oven, 1992 – see also, Chapter 6).

5.3 Effects upon the target

5.3.1 Introduction

In this section we analyse the effect of the host material as a consequence of ion bombardment. To this end, the target atoms are given a more active role in the mathematical description of penetration, instead of being simply 'sources' for deflection and energy loss.

A high-energy bombarding ion gives birth to a multiplicative effect known as a *collision cascade*, whereby primary recoil atoms can produce new recoils and so forth. This is illustrated in Figure 5.13.

Table 5.2. *The relative errors of KORAL, PRAL and KUBBIC with respect to TRIM85 for As into a-Si*

Moment		KORAL	PRAL	KUBBIC
$\langle z \rangle(E)$	Maximum error (%)	3.7	3.9	1.6
	Average error (%)	1.5	2.0	0.4
$\left[\langle \Delta z^2 \rangle(E) \right]^{1/2}$	Maximum error (%)	14.6	22.1	3.1
	Average error (%)	5.2	15.7	0.9
$\left[\langle \Delta x^2 \rangle(E) \right]^{1/2}$	Maximum error (%)	8.2	40.4	3.5
	Average error (%)	4.3	28.9	1.1

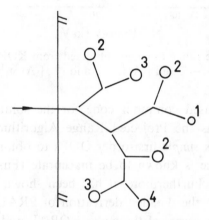

Fig. 5.13 Collision cascade: a bombarding particle 1 sets in motion primary knock-on atoms 2 which, in turn, set in motion a higher generation recoils, i.e. 3, 4, ... etc.

Several quantities within the cascade are of practical importance. For example, both the final and the initial location of recoil atoms relate to the interstitial and vacancy formation. These point defects are known to act as traps in semiconductor materials, thus introducing undesirable effects when doping by ion-implantation. The collision cascade here will be studied using exclusively the backward form of the BE.

5.3.2 The transport equation: the recoil term

The probability balance in equation (5.7) was derived assuming that only ions are counted by detector $\mathcal{D}$. However recoils can also be detected as illustrated in Figure 5.14.

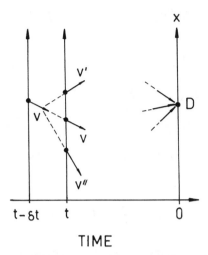

TIME

Fig. 5.14 The probability channel opened by a recoil particle. In the time interval $t - \delta t \rightarrow t$ the ion can traverse unscattered with velocity v or scatter with velocity v' inducing a recoil with velocity v''.

Equation (5.7) now becomes

$$
F^{\dagger}(r_0, v_0, t_0 - \delta t_0) = F^{\dagger}(r_0 + \delta t_0\, v_0, t_0) \left(1 - \delta t_0 \sum_{v_0'} P(v_0 \rightarrow v_0') \right)
$$

$$
+ \sum_{v_0'} \delta t_0\, P(v_0 \rightarrow v_0')[F^{\dagger}(r_0 + \delta t_0\, \hat{v}_0, v_0, t_0) + F^{\dagger}(r_0 + \delta t_0\, \hat{v}'', v'', t_0)] \qquad (5.80)
$$

$$
+ \delta t_0\, \mathscr{D}(r_0, v_0, t_0),
$$

where v'' stands for the recoil velocity, which is completely determined by momentum and energy conservation laws. Analogously, $\hat{v}''$ is an intermediate velocity, which will disappear as $\delta t_0 \rightarrow 0$.

We can transform (5.80) into an integro-differential equation by taking the limit as $\delta t_0 \rightarrow 0$, where omitting the subscript zero yields

$$
-\frac{\partial F^{\dagger}(r, v, t)}{\partial t} = (v \cdot \nabla_r)F^{\dagger}(r, v, t)
$$

$$
+ N|v| \int dv'\, \sigma(v, v')[F^{\dagger}(r, v, t) + F^{\dagger}(r, v', t) - F^{\dagger}(r, v'', t)] \qquad (5.81)
$$

$$
+ \mathscr{D}(r, v, t).
$$

Time-independent BE, including recoil term. Integrating (5.81) with respect to time gives

$$
\begin{aligned}
-(\boldsymbol{v} \cdot \nabla_{\boldsymbol{r}}) F^{\dagger}(\boldsymbol{r}, \boldsymbol{v}) = &-N|\mathbf{v}| \int d\boldsymbol{v}' \, \sigma(\boldsymbol{v}, \boldsymbol{v}') [F^{\dagger}(\boldsymbol{r}, \boldsymbol{v}) \\
&- F^{\dagger}(\boldsymbol{r}, \boldsymbol{v}') - F^{\dagger}(\boldsymbol{r}, \boldsymbol{v}'')] + \mathscr{D}(\boldsymbol{r}, \boldsymbol{v}),
\end{aligned}
\tag{5.82}
$$

where

$$
F^{\dagger}(\boldsymbol{r}, \boldsymbol{v}) = \int_{-\infty}^{+\infty} dt \, F^{\dagger}(\boldsymbol{r}, \boldsymbol{v}, t),
$$

$$
\mathscr{D}(\boldsymbol{r}, \boldsymbol{v}) = \int_{-\infty}^{+\infty} dt \, \mathscr{D}(\boldsymbol{r}, \boldsymbol{v}, t).
$$

In solving (5.73) it is necessary to specify a threshold energy E_0, which is required to displace an atom from its equilibrium site in the target. If this is not specified, not only is this unphysical but mathematical difficulties can arise with the number of recoiling atoms diverging.

5.3.3 Deposited energy distribution: equal mass cases

The distribution of elastically deposited energy will be denoted as $F_{D(1)}(\boldsymbol{v}, \boldsymbol{r})$. In the analysis electronic energy loss is ignored and the subscript one implies that ions and target are of the same species. The analysis will later be extended to the case in which ion and target are different. The deposited energy is calculated assuming the following scenario (Sigmund, 1972).

If within the collision cascade one particle hits a target atom and the transferred energy, T, becomes smaller than E_0 then one assumes that the energy T is being deposited within such a site. Otherwise the transferred energy will be carried away from the collision site. Analogously, if a moving particle's energy E becomes smaller than E_0 then E will be counted as having been deposited at that point. It is usual to choose E_0 as the potential barrier which the target atoms must surmount in order to escape from a lattice site. The atom, however, would recover such an energy after becoming free.

In order to determine $F_{D(1)}(\boldsymbol{v}, \boldsymbol{r})$ we shall use the steady-state backward BE and, as in equations (5.31), assume that

$$
F_{D(1)}(\boldsymbol{v}, \boldsymbol{r}) = F^{\dagger}(-\boldsymbol{r}, \boldsymbol{v}).
$$

Similarly to in (5.33),

$$
\mathscr{D}(\boldsymbol{r}, \boldsymbol{v}) = \delta(\boldsymbol{r}) \mathscr{D}(|\boldsymbol{v}|).
$$

Hence, the backward BE becomes

$$-(\mathbf{v} \cdot \nabla_r)F_{D(1)}(\mathbf{v}, \mathbf{r}) = N|\mathbf{v}| \int d\mathbf{v}' \sigma(\mathbf{v}, \mathbf{v}')[F_{D(1)}(\mathbf{v}', \mathbf{r}) + F_{D(1)}(\mathbf{v}'', \mathbf{r})$$
$$- F_{D(1)}(\mathbf{v}, \mathbf{r})] + \delta(\mathbf{r})\mathscr{D}(|\mathbf{v}|). \tag{5.83}$$

We proceed now in a similar manner to the case of projected range, by integrating over lateral position and introducing the energy, E, and direction cosine, η, as variables giving

$$\eta \frac{\partial F_{D(1)}}{\partial z} = N \int dT \, \mathbf{n}' \, d\mathbf{n}'' \, \sigma(E, T)f(\mathbf{n}, \mathbf{n}', \mathbf{n}'')[F_{D(1)}(E, \eta, z)$$
$$- F_{D(1)}(E - T, \eta', z) - F_{D(1)}(T, \eta'', z)] + \delta(z)\mathscr{D}(E), \tag{5.84}$$

where f also now depends on the direction $\mathbf{n}''$ of the recoil:

$$f(\mathbf{n}, \mathbf{n}', \mathbf{n}'') = \frac{1}{4\pi^2}\delta(\mathbf{n} \cdot \mathbf{n}' - \cos\psi')\delta(\mathbf{n} \cdot \mathbf{n}'' - \cos\psi'').$$

Here $\cos\psi'$ and $\cos\psi''$ stand for the scattering angle of the ion and that of the recoil in the laboratory frame of reference. The quantities η' and η'' represent the direction cosines of the particle and recoiling atom after scattering, respectively.

Spatial moments and Legendre expansion. Introducing the spatial moments and expanding in Legendre polynomials gives

$$n[lF_{D(1)l-1}^{(n-1)}(E) + (l+1)F_{D(1)l+1}^{(n-1)}(E)] = -(2l+1)N \int dT \, \sigma(E, T)$$
$$\times [F_{D(1)l}^{(n)}(E) - F_{D(1)l}^{(n)}(E - T)P_l(\cos\psi') - F_{D(1)l}^{(n)}(T)P_l(\cos\psi'')] \tag{5.85}$$
$$+ \delta_{n,0}\delta_{l,0}\mathscr{D}(E).$$

Normalization. As in the case of the range calculations we proceed by postulating

$$F_{D(1)l}^{(0)} = \delta_{l,0}E. \tag{5.86}$$

In this expression inelastic losses are ignored and all the deposited energy necessarily equals that of the incoming ion. Had we not neglected inelastic energy loss, the normalization (5.86) would have been replaced by

$$F_{D(1)l}^{(0)} = \delta_{l,0}v(E),$$

where $v(E) \leq E$. The equality holds only when inelastic energy loss becomes negligible small. This applies for low-energy or heavy ions.

Power scattering. For power scattering a solution to (5.85) can be found by assuming that

$$F^{(n)}_{D(1)l} = A^{(n)}_{D(1)l} E R^{n}_{(1)0}.$$

(5.87)

where $R_{(1)0} = E^{2m}/(NC_{(1)m})$. Here $C_{(1)m}$ denotes that the projectile and target are of the same species. This gives a recursive relationship for the dimensionless coefficients $A^{(n)}_{D(1)l}$,

$$A^{(n)}_{D(1)l} = \frac{n[lA^{(n-1)}_{D(1)l-1} + (l+1)A^{(n-1)}_{D(1)l+1}]}{(2l+1)[\mathfrak{I}^{(n)}_l(\gamma) - \mathcal{K}^{(n)}_l(\gamma)]},$$

(5.88)

with

$$A^{(0)}_{D(1)l} = \delta_{l,0},$$

$$\mathfrak{I}^{(n)}_l(\gamma) = \int_0^\gamma \frac{dx}{x^{1+m}} \left[1 - (1-x)^{1+2mn} P_l(\cos \psi') \right],$$

$$\mathcal{K}^{(n)}_l(\gamma) = \int_0^\gamma dx\, x^{2mn-m} P_l(\cos \psi'').$$

Here, $\gamma = 4m_1,m_2/(m_1+m_2)^2$ and from equation (2.41) with $x = T/E$,

$$\cos \psi' = (1-x)^{\frac{1}{2}} + \frac{m_1-m_2}{2m_1} \frac{x}{(1-x)^{\frac{1}{2}}},$$

(5.89)

$$\cos \psi'' = \frac{m_1+m_2}{2(m_1 m_2)^{\frac{1}{2}}} \sqrt{x}.$$

These equations provide the spatial moments of the deposited energy distribution.

5.3.4 Deposited energy distribution: non-equal mass cases

Having solved the transport equation for cases in which the ion and target atom have the same mass the extension to non-equal masses becomes straightforward. Let us now introduce $F_D(v,r)$ for the energy deposited by an ion different from those in the target. The target will be assumed to be elementary and homogeneous; hence the transport equation for such a function becomes

$$-(v \cdot \nabla_r)F_D(v,r) = \mathcal{D}(r,v)$$
$$+ N|v| \int dv'\, \sigma(v,v')[F_D(v',r) + F_{D(1)}(v'',r) - F_D(v,r)],$$

(5.90)

where the solution of the equal mass case $F_{D(1)}$ appears now in the place of the recoil term.

Spatial moments and Legendre expansion. Applying the same transformations as in the case of equal masses gives

$$
n[lF_{\mathrm{D}l-1}^{(n-1)}(E) + (l+1)F_{\mathrm{D}l+1}^{(n-1)}(E)] = -(2l+1)N \int \mathrm{d}T\, \sigma(E,T)
$$

$$
\times [F_{\mathrm{D}l}^{(n)}(E) - F_{\mathrm{D}l}^{(n)}(E-T)P_l(\cos\psi') - F_{\mathrm{D}(1)l}^{(n)}(T)P_l(\cos\psi'')] \qquad (5.91)
$$

$$
+ \delta_{n,0}\delta_{l,0}\mathscr{D}(E).
$$

Introducing the power-law scattering, one can define

$$
F_{\mathrm{D}l}^{(n)} = A_{\mathrm{D}l}^{(n)} E R_0^n. \qquad (5.92)
$$

Furthermore, in cases in which the same power applies both to ion–target and to target–target scattering, one can write

$$
A_{\mathrm{D}l}^{(n)} = \frac{n[lA_{\mathrm{D}l-1}^{(n-1)} + (l+1)A_{\mathrm{D}l+1}^{(n-1)}]}{(2l+1)\mathcal{K}_l^{(n)}(\gamma)} + \frac{A_{\mathrm{D}(1)l}^{(n)}(R_{(1)0}/R_0)^n}{\mathfrak{I}_l^{(n)}(\gamma)}, \qquad (5.93)
$$

where functions $\mathfrak{I}_l^{(n)}(\gamma)$ and $\mathcal{K}_l^{(n)}(\gamma)$ were defined before equations (5.89) and $A_{\mathrm{D}(1)l}^{(n)}$ refers to the constants associated with $F_{\mathrm{D}(1)l}^{(n)}$. Once the constants $A_{\mathrm{D}l}^{(n)}$ have been obtained, the spatial moments of the deposited energy distribution can be readily calculated,

$$
\langle z \rangle_{\mathrm{D}} = 3A_{\mathrm{D}1}^{(1)}P_1(\eta),
$$

$$
\langle z^2 \rangle_{\mathrm{D}} = A_{\mathrm{D}0}^{(2)} + 5A_{\mathrm{D}2}^{(2)}P_2(\eta).
$$

Assuming the same power law m for the ion–target and recoil–target interactions, we have calculated the mean depth and depth straggling of the deposited energy distribution as functions of the mass ratio and power $m = 1/4$, $1/3$ and $1/2$, respectively. The results are plotted in Figure 5.15. The mean depth and depth straggling of the deposited energy appear similar to those for range, see Figure 5.7. For large ion-to-target mass ratios, however, the mean depth of damage is smaller than the ion range.

5.3.5 *The distribution of displaced atoms*

The ultimate goal of this section is to have the recoil particles fully active within the process of transporting mass and energy within the collision cascade. In Section 5.3.4 we allowed recoils to be set into motion and to participate in distributing the ion energy within the target. Now, we determine the number of atoms in motion. This number is related to the observed damage which ion bombardment causes in the target material.

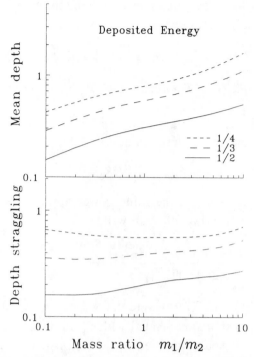

Fig. 5.15 Deposited energy. (a) Mean depth $\langle z \rangle$ and (b) depth straggling $\langle z^2 \rangle - \langle z \rangle^2$. Calculations are for power potentials $m = 1/2, 1/3$ and $1/4$. Inelastic energy loss is disregarded.

Such damage appears as a number of interstitials and vacancies, which can change electrical and mechanical properties of the target.

One can now define $F(E, \eta, z|E_0)\,\mathrm{d}z$ as the number of displaced atoms within the region $(z, z + \mathrm{d}z)$ produced by a projectile with energy E and direction cosine η, where E_0 stands for the displacement threshold. An infinite and homogeneous medium is again assumed, so that the backward transport equation can be used to obtain F.

The transport equation. It is assumed that $F(E, \eta, z|E_0)$ obeys the same backward BE as does the deposited energy, except that the detector term accounts for displaced atoms. We may therefore write

$$-\eta\frac{\partial F(E, \eta, z|E_0)}{\partial z} = -N\int\mathrm{d}T\,\mathrm{d}\boldsymbol{n}'\,\mathrm{d}\boldsymbol{n}''\,f(\boldsymbol{n}, \boldsymbol{n}', \boldsymbol{n}'')\sigma(E, T)[F(E, \eta, z|E_0)$$
$$-F(E - T, \eta', z|E_0) - F(T, \eta'', z|E_0)] + \mathscr{D}(z, E; E_0),$$

(5.94)

where writing $\mathscr{D}(z, E|E_0)$ explicitly denotes its dependence on E_0.

Spatial moments and Legendre expansion. Taking spatial moments and expanding in Legendre polynomials, we obtain

$$
\begin{aligned}
n[lF_{l-1}^{(n-1)}&(E|E_0) + (l+1)F_{l+1}^{(n-1)}(E|E_0)] \\
&= -(2l+1)N \int \mathrm{d}T\, \sigma(E,T) \left[F_l^{(n)}(E|E_0) \right. \\
&\quad \left. -F_l^{(n)}(E-T|E_0)P_l(\cos\psi') - F_l^{(n)}(T|E_0)P_l(\cos\psi'') \right] \\
&\quad + \delta_{n,0}\delta_{l,0}\mathscr{D}(E|E_0).
\end{aligned}
\tag{5.95}
$$

The detector term counts the amount of recoils with energy greater than E_0 that a unit dose of ions with energy E can produce after collision with target atoms. Therefore one can readily obtain

$$
\mathscr{D}(E|E_0) = N \int_{E \geq E_0} \mathrm{d}T\, \sigma(E,T).
$$

Using power scattering, one thus has

$$
\begin{aligned}
\mathscr{D}(E|E_0) &= \frac{NC_m}{E^m} \int_{E_0}^{E} \frac{\mathrm{d}T}{T^{1+m}} \\
&= \frac{NC_m}{mE^m E_0^m} [1 - (E_0/E)^m]\Theta(E-E_0);
\end{aligned}
\tag{5.96}
$$

therefore,

$$
\begin{aligned}
n[lF_{l-1}^{(n-1)}&(E|E_0) + (l+1)F_{l+1}^{(n-1)}(E|E_0)] \\
&= -(2l+1)\frac{NC_m}{E^m} \int_0^E \frac{\mathrm{d}T}{T^{1+m}} \left[F_l^{(n)}(E|E_0) \right. \\
&\quad \left. - F_l^{(n)}(E-T|E_0)P_l(\cos\psi') - F_l^{(n)}(T|E_0)P_l(\cos\psi'') \right] \\
&\quad + \delta_{n,0}\delta_{l,0}\frac{NC_m}{mE^m E_0^m} [1 - (E_0/E)^m]\Theta(E-E_0).
\end{aligned}
\tag{5.97}
$$

Here Θ is the Heaviside step function

$$
\Theta = \begin{cases} 0 & E < E_0 \\ 1 & E > E_0. \end{cases}
$$

Asymptotic solution. Here, in contrast to the solution of deposited energy or range, one cannot assume a power of the energy as a solution since $F_l^{(n)}(E|E_0) = 0$ for $E < E_0$. Instead we obtain an asymptotic solution valid in the limit $E \gg E_0$. The idea behind the method which we shall follow to obtain asymptotic solutions is that of Morse and Feschbach (1953). Suppose that we have a certain function, $F(E)$, which is known to be zero for $E < E_0$

and $F(E) = \kappa E^{\nu}$ for $E \gg E_0$, then we can introduce a variable $u = \ln(E/E_0)$, so that

$$F(u) = \begin{cases} \kappa E_0^{\nu} \exp(\nu u) \; ; \; \text{if } u > 0 \\ \qquad\qquad 0; \; \text{otherwise.} \end{cases} \tag{5.98}$$

Taking the Laplace transform of $F(u)$ one obtains

$$\bar{F}(s) = \kappa E_0^{\nu} \int_0^{+\infty} du \, \exp[-(s - \nu)u],$$
$$= \frac{\kappa E_0^{\nu}}{s - \nu}, \tag{5.99}$$

which shows that ν equals the right-most real pole of $\bar{F}(s)$ and that κ becomes in the case of simple poles

$$\kappa = \lim_{s \to \nu} \frac{(s - \nu)\bar{F}(s)}{E_0^{\nu}}.$$

Let us apply such a method to equation (5.97) for the case of $n = 0$, $l = 0$:

$$\int_0^E \frac{dT}{T^{m+1}} [F_0^{(0)}(E|E_0) - F_0^{(0)}(E - T|E_0) - F_{(0)}(T|E_0)]$$
$$= \frac{1 - (E_0/E)^m}{m E_0^m} \Theta(E - E_0). \tag{5.100}$$

Transforming equation (5.100) gives

$$\int_0^E dT \, \{F_0^{(0)}(E)T^{-1-m} - F_0^{(0)}(T)[(E - T)^{-1-m} + T^{-1-m}]\}$$
$$= \frac{1 - (E_0/E)^m}{m E_0^m} \Theta(E - E_0); \tag{5.101}$$

then one can write

$$\int_0^1 dx \, \{F_0^{(0)}(E)x^{-1-m} - F_0^{(0)}(Ex)[(1 - x)^{-1-m} + x^{-1-m}]\}$$
$$= \frac{(E/E_0)^m - 1}{m} \Theta(E - E_0), \tag{5.102}$$

by introducing the variable $u = \ln(E/E_0)$ we thus have

$$F_0^{(0)}(E) \to F_0^{(0)}(u),$$

$$F_0^{(0)}(Ex) \to F_0^{(0)}(u + t),$$

with

$$t = \ln x.$$

Hence, after multiplying equation (5.102) by $\exp(-su)$ and integrating over u from zero to infinity, we obtain

$$\bar{F}_0^{(0)}(s)\Lambda_0(s) = \frac{1}{s(s-m)}, \tag{5.103}$$

where

$$\bar{F}_0^{(0)}(s) = \int_0^{+\infty} du \exp(-su)\, F_0^{(0)}(u),$$

$$\Lambda_0(s) = \int_0^1 ds\,\{x^{-1-m} - [(1-x)^{-1-m} + x^{-1-m}]x^s\}$$
$$= B(-m,1) - B(s-m,1) - B(s+1,-m), \tag{5.104}$$

$B(x,y)$ being the beta function (Abramowitz and Stegun, 1970). The subscript zero in function Λ implies that it corresponds to the case of $l = 0$ in the Legendre expansion.

Alternatively one can express the function $\Lambda_0(s)$ in terms of the gamma function as

$$\Lambda_0(s) = \frac{\Gamma(-m)\Gamma(1)}{\Gamma(1-m)} - \frac{\Gamma(s-m)\Gamma(1)}{\Gamma(1-m+s)} - \frac{\Gamma(1+s)\Gamma(-m)}{\Gamma(1-m+s)}. \tag{5.105}$$

From equation (5.103)

$$\bar{F}_0^{(0)}(s) = \frac{1}{\Lambda_0(s)}\frac{1}{s(s-m)}.$$

The right-most pole of the expression above occurs when $\Lambda_0(s) = 0$ i.e. $s = 1$ if $m < 1$. In addition, from (5.99), when $v = 1$,

$$\kappa = \frac{1}{(1-m)\Lambda_0'(1)E_0}.$$

According to equation (5.105), one can readily obtain

$$\Lambda_0'(1) \simeq \frac{\psi(1) - \psi(1-m)}{m(1-m)},$$

where $\psi(x) = \ln \Gamma(x)$. Given that $m < 1$ one can approximate

$$\psi(1) - \psi(1-m) \simeq m\psi'(1) \simeq \frac{m\pi^2}{6};$$

hence,

$$F_0^{(0)}(E|E_0) \simeq \frac{6}{\pi^2}\frac{E}{E_0}. \tag{5.106}$$

This result indicates that the total number of recoils produced by a projectile of energy E becomes, asymptotically, a linear function of the bombarding energy and inversely proportional to the displacement energy E_0. Notice also that the total number of recoils does not depend on the power m of the scattering cross-section. Note also that for $n = 0$, $l > 0$ equation (5.95) is satisfied by $F_l^{(0)} = 0$.

Higher moments: the equal mass case. In solving equation (5.100) for the higher moments it is convenient to assume an asymptotic form for the left-hand side of (5.97):

$$n[lF_{l-1}^{(n-1)}(E|E_0) + (l+1)F_{l+1}^{(n-1)}(E|E_0)] \sim \kappa E^v \Theta(E - E_0),$$

thus (5.97) becomes

$$\kappa E^v \Theta(E - E_0) = \frac{NC_m}{E^m} \int_0^E \frac{dT}{T^{1+m}} \left\{ F_l^{(n)}(E|E_0) \right.$$
$$\left. -F_l^{(n)}(E - T|E_0)P_l\left[(1 - T/E)^{\frac{1}{2}}\right] - F_l^{(n)}(T|E_0)P_l\left[(T/E)^{\frac{1}{2}}\right] \right\}, \tag{5.107}$$

where we have expressed $\cos \psi'$ and $\cos \psi''$ in terms of relative energy loss according to equations (5.89), for $m_1 = m_2$.

Following the method introduced before, one obtains

$$\bar{F}_l^{(n)}(s) \sim \frac{E_0^{v+2m}}{(s - v - 2m)\Lambda_l(s)}, \tag{5.108}$$

where

$$\Lambda_l(s) = \int_0^1 ds \left\{ x^{-1-m} - x^s P_l(\sqrt{x})[(1 - x)^{-1-m} + x^{-1-m}] \right\} \tag{5.109}$$

is a function whose largest real zero cannot be greater than that of $\Lambda_0(s)$.

From (5.106) we obtain

$$\Lambda_l(s) = \Lambda_0(s) + \Delta_l(s), \tag{5.110}$$

where

$$\Delta_l(s) = \int_0^1 dx \, x^s \left[1 - P_l(\sqrt{x}) \right][x^{-1-m} + (1 - x)^{-1-m}].$$

Whence, because $|P_l(x)| \leq 1$, the last integral must be non-negative for all real s, we can conclude that the right-most zero of $\Lambda_l(s)$ must necessarily be smaller than that of $\Lambda_0(s)$. Therefore, if $v \geq 1$, the right-most pole of $\bar{F}_l^{(n)}(s)$ will be $s = v + 2m$, hence

$$F_l^{(n)}(E) \sim \frac{E^{v+2m}}{\Lambda_l(v + 2m)}. \tag{5.111}$$

The result above could have been obtained by following the procedure used when calculating deposited energy, for which we assumed that $E_0 = 0$ and approximated $F_{Dl}^{(n)}(E)$ by a certain power of E.

The cases of $F_l^{(n)}(E)$ with $n \geq 1$ are derived exactly as before. Starting with $n = 1$ we have

$$
\begin{aligned}
F_0^{(0)}(E|E_0) = -3\frac{NC_m}{E^m} \int_0^E \frac{dT}{T^{1+m}} &\left[F_1^{(1)}(E|E_0) \right. \\
&\left. - F_1^{(1)}(E - T|E_0)P_1(\cos\psi') - F_1^{(1)}(T|E_0)P_1(\cos\psi'') \right],
\end{aligned}
\tag{5.112}
$$

where, as seen already, $F_0^{(0)}(E|E_0) \sim E$ and $F_2^{(0)} = 0$. Thus

$$
F_1^{(1)} \simeq E^{1+2m}
$$

and similar forms will also occur with higher moments. All the moments for the deposited energy distribution are proportional to those of the deposited energy distribution so

$$
F(E, \eta, z|E_0) \sim F_D(E, \eta, z),
\tag{5.113}
$$

where the proportionality factor can be readily obtained from the fact that F_D is normalised with respect to E whereas the normalisation value of the recoil distribution is given by equation (5.106):

$$
F(E, \eta, z|E_0) \simeq \frac{6}{\pi^2} \frac{F_D(E, \eta, z)}{E_0}.
\tag{5.114}
$$

As an example, Figure 5.16 shows the damage left in a silicon target by 8×10^{16} ions cm^{-2} 100 keV Si ions (Gibbons and Christel, 1984). The damage is observed as the enhancement in the number of protons back-scattered from the target as a function of depth. The incoming proton beam is directed along a channelling direction so for an undamaged crystal the back-scattering yield would be low. However, because Si atoms have been displaced from their sites, the back-scattering yield is proportional to the damage. In Figure 5.16 is also plotted the deposited-energy distribution for 100 keV Si on Si, using the first-order Gaussian approximation. In this case we use $\varepsilon = 2.3$ ($m = \frac{1}{2}$) as the appropriate power and ignore inelastic losses. Despite the fact that the inelastic stopping of Si on Si is approximately 30% that of nuclear stopping at 100 keV, the deposited energy distribution reproduces relatively well the shape of the back-scattering yield.

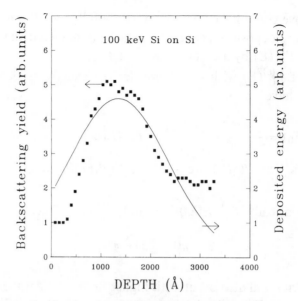

Fig. 5.16 Damage distribution in an Si target after 8×10^{16} ions cm^{-2}, 100 keV Si ion bombardments. Experiments: full squares, which represent the yield of back-scattered protons. Theory: continuous line.

5.4 Sputtering

The ejection of atoms from the target surface, known as sputtering, has become central to many technological developments. Quantitatively, sputtering can be characterized by the so-called sputtering yield. This is defined as the mean number of ejected atoms per incident ion.

Transport theory has proven to be a powerful tool in providing simple and general expressions for various observables in the sputtering processes (Sigmund, 1969, Thompson, 1968). The sputtering yield, in particular, can be reasonably well predicted by a simple expression resulting from first-principles transport calculations.

In this section we closely follow Sigmund's (1969) model of sputtering. Accordingly, we shall calculate the number of recoil particles that can penetrate a certain reference plane with energy greater than U. That is, one can obtain a function $H(E, \eta, z)$ for the flux of recoils capable of overcoming a planar potential barrier of height U, located at a depth z within the target. The ions are assumed to arrive at $z = 0$ with energy E and direction cosine η. The sputtering yield is obtained as

$$Y = H(E, \eta, 0)$$

As in the preceding section, we can utilize the backward BE after assuming an infinite homogenous medium. The detector term in this case becomes

$$\mathscr{D}(r,v) = \delta(r)\begin{cases} |v_z| \; ; & \text{if } \frac{1}{2}m_2v_z^2 > U \text{ and } v_z < 0 \\ 0 \; ; & \text{otherwise.} \end{cases} \tag{5.115}$$

Such an expression represents a distribution of detectors at the origin count-ing the flux of atoms moving towards the negative direction of the z axis with kinetic energy associated with such a direction greater than U.

Writing the detector term in terms of energy, after expansion of (5.94) in Legendre polynomials and spatial moments, one has

$$n[lH_{l-1}^{(n-1)}(E) + (l+1)H_{l+1}^{(n-1)}(E)] = -(2l+1)N$$
$$\times \int dT \, d\sigma(E,T)\,[H_l^{(n)}(E) - H_l^{(n)}(E-T)P_l(\cos\psi') \tag{5.116}$$
$$- H_l^{(n)}(T)P_l(\cos\psi'')] + (2l+1)Q_l(E)\delta_{n,0},$$

where

$$Q_l(E) = \begin{cases} \frac{1}{2}\int_{-1}^{-(U/E)^{\frac{1}{2}}} dx\,|x|P_l(x); & \text{if } E > U, \\ 0; & \text{otherwise.} \end{cases}$$

For example, for $E > U$ one obtains

$$Q_0(E) = \frac{1-U/E}{4},$$
$$Q_1(E) = \frac{1-(U/E)^{3/2}}{6} \tag{5.117}$$

and so on.

Again, using the power approximation for the scattering cross-section, equation (5.116) reads

$$n[lH_{l-1}^{(n-1)}(E) + (l+1)H_{l+1}^{(n-1)}(E)] = -(2l+1)\frac{NC_m}{E^m}$$
$$\times \int \frac{dT}{T^{1+m}} \times [H_l^{(n)}(E) - H_l^{(n)}(E-T)P_l(\cos\psi') \tag{5.118}$$
$$- H_l^{(n)}(T)P_l(\cos\psi'')] + (2l+1)Q_l(E)\delta_{n,0}.$$

In the cases of the zeroth-order moments one has

$$Q_l(E) = \frac{NC_m}{E^m}\int \frac{dT}{T^{1+m}}[H_l^{(0)}(E) -$$
$$H_l^{(0)}(E-T)P_l(\cos\psi') - H_l^{(0)}(T)P_l(\cos\psi'')]. \tag{5.119}$$

For simplicity in evaluating the integrals we will again assume that $m_1 = m_2$.

We can now proceed in a similar manner to the case of deposited energy. Introducing, first, the variable $u = \ln(E/U)$ we can then multiply both sides of equation (5.119) by $\exp(-us)$ and integrate over u. In this way one can obtain an equation for the Laplace transform of function $H_l^{(0)}(E)$. Finally,

$$\bar{H}_l^{(0)}(s) = \frac{U^{2m}}{NC_m} \frac{f_l(s)}{2(s - 2m)\Lambda_l(s)},$$ (5.120)

with

$$f_l(s) = \frac{1}{2} \int_0^1 dx\, x^{(s-2m)} P_l(-\sqrt{x}),$$

where

$$f_0(s) = \frac{1}{2(s + 1 - 2m)},$$

$$f_1(s) = \frac{1}{2(s + 3/2 - 2m)},$$

etc.

In general, given that $P_l(x)$ is a polynomial of degree l, one can easily see that real poles of $f_l(s)$ are all smaller than $2m - 1$. Therefore, according to our discussion about the zeros of $\Lambda_l(s)$ we can readily conclude that, so long as $l > 0$,

$$\frac{f_l(s)}{(s - 1/2 - 2m)\Lambda_l(s)}$$

cannot have real poles greater than or equal to unity. Consequently, the asymptotic solutions in cases in which $l > 0$ are smaller than that of $l = 0$.

It is now possible to proceed in much the same manner as with the recoil distribution. We can obtain

$$H_0^{(0)}(E) = \frac{U^{2m}}{NC_m} \frac{f_0(1)}{2(1 - 2m)\Lambda_0'(1)} \frac{E}{U},$$ (5.121)

and if we also assume that $H_l^{(0)}(E)$ is small for $l > 0$ the flux of atoms H must be proportional to the deposited energy distribution F_D,

$$H(E, \eta, z) \propto F_D(E, \eta, z).$$

The proportionality factor can be obtained from the zeroth-order spatial moments solution, equation (5.121), the result being

$$H(E, \eta, z) = \frac{U^{2m}}{NC_m} \frac{f_0(1)}{2(1 - 2m)\Lambda_0'(1)} \frac{F_D(E, \eta, z)}{U}.$$

Then, evaluating $H(E, \eta, 0)$, we have

$$Y = \frac{U^{2m}}{NC_m} \frac{f_0(1)}{2(1 - 2m)\Lambda_0'(1)} \frac{F_D(E, \eta, z = 0)}{U}.$$ (5.122)

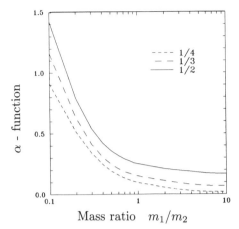

Fig. 5.17 The function α obtained using the Gaussian approximation without including inelastic energy loss for three different values of m, as a function of m_1/m_2.

On the other hand, one can identify $U^{2m}/(NC_m)$ with the mean range of a particle with energy U, i.e., $\Lambda = U^{2m}/(NC_m)$, so that

$$Y \simeq \frac{3}{4\pi^2} \frac{\Lambda F_D(E, \eta, z = 0)}{U}. \tag{5.123}$$

The sputtering yield results from the energy deposited within depth Λ at the surface, divided by the height U of the potential barrier.

Expression (5.123) can be further simplified by approximating the deposited energy as

$$F_D(E, \eta, z = 0) = \alpha(E, \eta, m_1/m_2)NS_n(E), \tag{5.124}$$

where S_n is the nuclear stopping power and α a dimensionless function that depends on the bombarding energy, E, the angle of incidence, η, and the mass ratio, m_1/m_2. Equation (5.123) thus becomes

$$Y = \frac{3}{4\pi^2}\alpha\frac{\Lambda NS_n(E)}{U}. \tag{5.125}$$

Using the spatial moments for the deposited energy calculated in the previous section we can obtain the function α using the Gaussian approximation

$$\alpha = \frac{E}{NS_n(E)((2\pi\langle\Delta z^2\rangle)^{\frac{1}{2}}} \exp\left(-\frac{\langle z\rangle^2}{2\langle\Delta z^2\rangle}\right). \tag{5.126}$$

Since $NS_n(E) = \gamma^{1-m}E/(1-m)R_0$, one thus has

$$\alpha = \frac{1-m}{\gamma^{1-m}}\left(\frac{\langle\Delta z^2\rangle}{2\pi R_0^2}\right)^{\frac{1}{2}} \exp\left(-\frac{\langle z\rangle^2}{2\langle\Delta z^2\rangle}\right). \tag{5.127}$$

It is worthwhile noting that α represents the fraction of the mean energy loss per unit path of the incoming ion that is *actually* deposited in the target surface. Such a fraction is not necessarily unity, since not only is the energy carried away from the surface by the recoils but also the ion may pass through the surface at least once again, thus depositing additional energy on the target surface. Figure 5.17 shows that α monotonically decreases with increasing ion mass. For light ions it becomes greater than unity, indicating that ion may pass through the surface several times (a limitation of the infinite-target approximation). At large ion-to-target mass ratios, α becomes smaller than unity and asymptotically approaches the infinite heavy-ion value. Then, the deposited energy is entirely determined by the recoil atoms in the target.

Analogously, since Λ accounts for the range of the low-energy recoils, one can approximate (Sigmund, 1969)

$$\Lambda \approx \frac{1}{NC_0},\qquad(5.128)$$

with $C_0 = 1.81$ $\mathring{A}^2$ (Sigmund, 1972).

This gives

$$Y\ (\text{at/ions}) = \frac{0.042\alpha S_n(E)\ (\text{V}\ \mathring{A}^2)}{U\ (\text{eV})}.\qquad(5.129)$$

A comparisons of equation (5.129) with experimental data is shown in Figure 5.18. Despite the various approximations entering such an equation the agreement with experiments is good.

Deviations from theoretical predictions occurring particularly with high-sputtering-yield targets have been attributed to *non-linear* effects within the collision cascade. When the binding energy is small and ion and target masses become both, say, greater than 100 a.m.u. the collision cascades are expected to be particularly 'dense', so that either collision between moving particles or 'overlap' between sub-cascades is likely and the linear description thus fails. For example, a large sputtering yield is observed in the case of Xe on Au around the maximum of the nuclear stopping at which, as we can see in Figure 5.18, the predictions of equation (5.129) are nearly a factor of four down from the experiments.

Energy-distribution of ejected particles. By following the procedure previously employed to obtain the sputtering yield, one can obtain the energy spectrum of the ejected atoms. This can be achieved by modifying the detector term

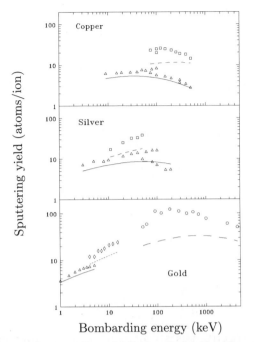

Fig. 5.18 Experimental sputtering yields (symbols) for Cu, Ag, and Au targets compared with calculated sputtering yields (lines) △ Ar, □ Xe, ○ Au and ◇ Hg bombardment. Note that the heavier the ion, the less good the agreement with experiment.

in (5.115) so as to contain an energy window, i.e.

$$\mathcal{D}(\boldsymbol{r}, \boldsymbol{v}) = \delta(\boldsymbol{r}) \left(\frac{1}{2} m_2 v^2 - U - E_0 \right) \begin{cases} |v_z| & \text{if } \frac{1}{2} m_2 v_z^2 > U \text{ and } v_z < 0 \\ 0 & \text{otherwise.} \end{cases} \quad (5.130)$$

Here E_0 is the energy of the ejected atoms after their having overcome the surface potential. After expansion in Legendre polynomials and taking spatial moments one obtains

$$n \left[l G_{l-1}^{(n-1)}(E|E_0) + (l+1) G_{l+1}^{(n-1)}(E|E_0) \right]$$
$$= \left[-(2l+1) \frac{NC_m}{E^m} \int_0^E \frac{\mathrm{d}T}{T^{1+m}} G_l^{(n)}(E|E_0) \right.$$
$$\left. - G_l^{(n)}(E-T|E_0) P_l(\cos \psi) - G_l^{(n)}(T|E_0) P_l(\cos \psi') \right]$$
$$+ \delta_{n,0} Q_l(E) \delta(E - E_0 - U). \quad (5.131)$$

The function $Q_l(E)$ is defined by equation (5.117), and $G(E, \eta, x|E_0)$ represents the number of atoms passing x with energy $(E_0, \mathrm{d}E_0)$ as a consequence of an ion starting at $x = 0$ with energy E and direction cosine η. $G(E, \eta, 0|E_0)$, therefore, will account for the energy spectra of the ejected atoms integrated over all directions after subtraction of the binding energy U for the atoms in the target.

We can now proceed in a similar manner to that with the sputtering yield and obtain the asymptotic solutions of equation (5.131), for $E \gg U$. The result being

$$G_0^{(0)}(E|E_0) \simeq \frac{3}{2\pi^2} \frac{E_0 U}{(E_0 + U)^3} \frac{E}{NC_0 U}.$$

One can easily verify that $G_l^{(0)} \propto (E/U)^\nu$ with $\nu < 1$, so that the leading term in the Legendre expansion of the zeroth spatial moment of function G will be that of $l = 0$. Similarly, the results for $n > 0$ can be obtained giving

$$G(E, \eta, x|E_0) \propto F_D(E, \eta, x).$$

Thus, the energy spectrum for the ejected atoms becomes

$$\frac{dN}{dE_0} = \frac{2E_0 U}{(E_0 + U)^3} Y, \tag{5.132}$$

indicating that the spectrum is strongly peaked at lower energies, with a maximum at $E_0 = U/2$ and decaying as E_0^2 at higher energies. This result indicates the importance that low energy recoils have in the sputtering of surfaces.

The transport theory formulation has provided general and simple expressions but it will be shown in Chapter 8 that the ejection of atoms from surfaces comprises a number of mechanisms that are not well modelled by the approximations employed in this chapter. The peak in the energy distribution of ejected particles is not always at $E_0 = U/2$ for example for crystalline covalent materials. A summary of results of transport theory applied to sputtering has been given by Sigmund (1981).

5.5 Mixing

5.5.1 Introduction

On their passage through the target the ions may disturb the host material, displacing target atoms within the collision cascade. If the composition of the target varies with depth such as in layered solids, the distribution of target species is modified by ion bombardment.

Figure 5.19 shows a target containing a thin layer of 'B'-type atoms on top of 'A'-type substrate *before* and *after* ion-beam bombardment. As the dose increases the interface becomes less sharp, the surface recedes due

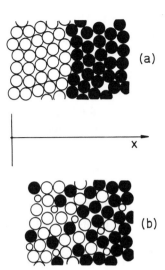

Fig. 5.19 Atomistic view of a thin 'B'-atom layer (open circles) atop of an 'A' substrate (full circles). (a) Before ion-bombardment and, (b) After ion bombardment. Ions are shown as small open circles.

to sputtering and bombarding atoms become implanted within the target. Strongly motivated by the possibility of fabricating non-equilibrium alloys and because of the problems of interpreting depth profiles in surface analysis, the phenomena of ion-beam mixing has been investigated in detail in recent years.

Apart from relocations produced by the ion or generation of more recoils, diffusion may also contribute to the migration of atoms within the target. Because the damage left by ion-bombardment enhances diffusion, atomic migration may therefore increase with the ion dose.

This phenomenon, known as radiation-enhanced diffusion (RED), is purposely avoided in this chapter since the collision cascade and this type of mass transport are rather indirectly connected. RED is a post-bombardment effect in the sense that it takes place on a much longer time-scale than that of the collision cascade. In the same manner, chemically driven diffusion will be also disregarded.

5.5.2 Modelling ion-beam atomic mixing

Mixing, contrary to the various ion–target processes modelled so far in this chapter, is a non-linear problem. In fact, in calculating range or recoil distribution one presumes an undisturbed target. However, in order to obtain the distance that atoms are being displaced within a collision cascade one has to monitor the changes to the composition profile after successive hits. Such information cannot be obtained from the BE under the linear approximation. So, unless a non-linear form of the BE is used, mixing can be solved only in an approximate manner. To overcome this problem it is assumed that the linear BE can be applied within infinitesimal variation of fluence. Non-linearity enters only later, at the time of integrating over bombarding dose. Accordingly one can write a non-linear equation for the evolution of concentration with fluence, while keeping the linear approach to calculate the 'instantaneous' values of several quantities entering such an equation.

This model of mixing is not amenable to analytical solution except for a few, over-simplified, cases. More realistic treatment, even though containing many simplifying assumptions, can be afforded only by numerical means (see for instance Wadsworth *et al.* (1990), Kim *et al.* (1988) and Chou and Ghoniem (1986)).

Basic mixing equations. Suppose that we have a target containing a number of atomic species $i = 1,\ldots,s$ initially arranged according to a certain concentration distribution. The species of the incoming ion will be denoted by $i = 1$. We assume a one-dimensional target located along the positive X axis and that the ion bombardment begins at $t = 0$ with a current density I ions cm^{-2} s^{-1}. In the formulation that follows we will work with the independent variables t and X rather than ϕ and X, where ϕ is the ion dose, usually linearly related to t. If $J_i(X)$ represents the flux-density of atoms i at depth X then a normal conservation of mass argument would lead to

$$\frac{\partial}{\partial t}N_i(X,t) = -\frac{\partial}{\partial X}J_i(X,t). \qquad (5.133)$$

This has the boundary condition

$$J_i(0,t) = -I(Y_i + R\delta_{i1}),$$

where Y_i is the sputtering yield of species i, R is the reflection coefficient and δ_{ij} is the Kronecker delta function.

Equation (5.133) is not especially useful in the form given because the distance X is measured from some absolute point in space. As the target is bombarded it can swell to accommodate the arriving ions. The surface can also recede due to sputtering. For the time being we ignore the effect of surface erosion and instead introduce a new co-ordinate x' which redefines depth to take into account target swelling

$$x' = X - \int_0^t k(X,t')\,dt',$$ (5.134)

where

$$k(X,t) = \sum v_i J_i(X,t),$$ (5.135)

v_i being the volume of atom i chosen so that $\sum \tau_i N_i(X,0) = 1$. This way of measuring depth will either 'compress' or 'expand' the depth scale so that the volume occupied by each atomic species is conserved.

Let $n_i(x',t)$ be the number density written in terms of x', then

$$\tau_i n_i(x',t)\,dx' = \tau_i N_i(X,t)\,dX.$$

Using this in (5.134) we have

$$n_i(x',t) = \frac{N_i(X,t)}{\left[1 - \int_0^t dt'\,\partial k(X,t')/\partial X\right]}.$$ (5.136)

From (5.133) and (5.135)

$$n_i(x',t) = \frac{N_i(X,t)}{\sum_v \tau_i N_j(X,t)}.$$ (5.137)

Thus

$$\sum_i \tau_i n_i(x',t) = 1.$$ (5.138)

If, now, time changes by a small amount δt keeping X' fixed, from (5.138), then we have

$$\delta X = \frac{k(X,t)\,\delta t}{\sum_j \tau_j N_j(X,t)}.$$ (5.139)

Now, expand $n_i(x',t+\delta t)$ in a Taylor series, keeping terms up to linear ones, to give, after some algebra

$$\frac{\partial n_i(x',t)}{\partial t} = -\frac{\partial}{\partial x'}\left[J_i(x') - n_i(x',t)k(x',t)\right].$$ (5.140)

This equation contains an additional 'current' term $n_i(x',t)k(x',t)$ directed towards the surface. At the surface, $X = 0$ so the position of the surface x'_s is given by

$$x'_s = -\int_0^t dt'k(0,t') = \int_0^t dt'\,U(t'),$$

where U is the recession speed of the surface. This relationship can be used to renormalise the depth scale by subtracting the instantaneous position of the surface. Redefining $x = x' - x_s$, equation (5.140) can be written

$$\frac{\partial}{\partial t} n_i(x, t) = -\frac{\partial}{\partial x} J_i^{(\text{eff})}(x, t),$$ (5.141)

where

$$J_i^{(\text{eff})}(x, t) = J_j(x, t) - [k(x, t) + U] n_i(x, t).$$ (5.142)

The mass current J_i can be approximated by a sum of terms corresponding to the different mechanics, which can produce redistribution of atoms, such as direction and recoil relocation, diffusion in the cascade and sputtering.

Direct ion and recoil relocation. Here we introduce a function $F_i(x, z)$ such that $n_i F_i(x, z) \, dx \, dz$ is the cross-section for relocation i.e. the mean number of i atoms travelling from the interval (x, dx) to $(x + z, dz)$ due to one incident ion. The cascade current due to these relocations is given by

$$\begin{aligned} J_i^{(\text{cas})}(x, t) = {} & I \int_0^x d\eta \, n_i(\eta, t) \int_{x-\eta}^\infty dz \, F_i(\eta, z) \\ & - I \int_x^\infty d\eta \, n_i(\eta, t) \int_{-\infty}^{x-\eta} dz \, F_i(\eta, z) \\ & + \delta_{i1} I \int_x^\infty d\eta \, F_R(\eta). \end{aligned}$$ (5.143)

The first term on the right-hand side of equation (5.143) represents the flux of atoms going from left to right through a plane located at depth x. The second term accounts for the flux of atoms in the opposite direction. The last term is zero if $i \neq 1$ and here F_R is the range distribution of the ions.

The function $F_i(x, z)$ is difficult to calculate because it will change as a function of ion dose (and thus time) as the composition changes. Sigmund and Gras-Marti (1981) have calculated analytic expressions for $F_i(x, z)$ in a few special cases for a dilute impurity using power law potentials. The derivation of these expressions can be summarised as follows. First the relocation function is related to the density of moving particles within the cascade, $G(x, E/E_0, \Omega_0)$,

$$\begin{aligned} F_i^{(j)}(x, z) = {} & \int d\Omega_0 \, dE_0 \, G_j(x, E/E_0, \Omega_0) |\cos \theta_0|^{-1} \\ & \times \int dE' \, d\Omega' \, \sigma_{j,i}(E_0, \Omega_0 \to E', \Omega') F_{Ri}(E', \Omega'|z), \end{aligned}$$ (5.144)

where E_0 and Ω_0 are the instantaneous energy and direction of the moving particle, θ_0 is the angle between Ω_0 and the x axis and E is the incoming ion energy. The function $F_{Ri}(E', \Omega'|z)$ is the range distribution of an i atom with energy E' and direction Ω'. The subscript j identifies the moving particle. Hence $\sigma_{j,i}(E_0, \Omega_0 \rightarrow E', \Omega')$ represents the cross-section for a j particle to set in motion an i atom within (E', dE') and $\Omega', d\Omega'$). The $|\cos\theta_0|^{-1}$ term is necessary to convert depth into path along direction Ω_0.

The straight-line ion trajectory. Using the straight-line approximation for the ion trajectory

$$G_1(x, E|E_0, \Omega_0) \simeq \frac{1}{2\pi} \delta(E_0 - E(x)) \delta(\cos\theta_0 - 1),$$

where $E(x)$ is the mean energy of the ions at depth x. Also approximating

$$F_{Ri}(E', \Omega'|z) \simeq \delta(z - A(E')^\alpha \cos\theta'), \tag{5.145}$$

where θ' is the angle between Ω' and the x axis, then using a power law potential with exponent m to calculate the impurity cross-section gives

$$F_i^{(1)}(x, z) \simeq \frac{2}{2\alpha+1} \frac{B(x)}{z_{max}} (z_{max}/z)^{1+m'} \text{ for } 0 \le z \le z_{max}. \tag{5.146}$$

Here $m' = 2m/(2\alpha+1)$, $z_{max} = A|\gamma_{1i}E(x)|^\alpha$, the maximum range of the knock-off impurity, $\gamma_{1i} = 4m_1 m_i/(m_1 + m_i)^2$ and $B(x) = C_m/|\gamma_{1i}E^2(x)|^m$.

Isotropic ion distribution. When the distribution of moving ions is isotropic

$$G_1(x, E|E_0, \Omega_0) \simeq \frac{1}{4\pi} \frac{R(E)}{R_p(E)} \delta(E_0 - E(x)),$$

where $R(E)$ is the range along the path and R_p is the projected range of the ions. Using (5.144) gives

$$F_i^{(1)}(x, z) \simeq \frac{R(E)}{2R_p(E)} \frac{B(x)}{z_{max}} (z_{max}/z)^{1+m/\alpha} \text{ for } |z| \le z_{max}. \tag{5.147}$$

Isotropic recoil distribution. For an isotropic distribution of recoils

$$G_2(x, E|E_0, \Omega_0) \simeq \frac{\Gamma}{4\pi N} \frac{F_D(x)|\cos\theta_0|}{E_0 \, s_{22}(E_0)}, \tag{5.148}$$

where s_{22} is the stopping cross-section of target recoils within their own matrix, N the atomic number density, Γ is a dimensionless coefficient approximately equal to unity and F_D is the energy deposited per unit depth in elastic collisions. This density distribution gives

$$F_i^{(2)}(x, z) \simeq \Gamma\xi_{2i} \frac{F_D(x, E)}{NE_c R_c} (R_c/|z|)^{1+1/\alpha}, \text{ for } |z| > R_c, \tag{5.149}$$

where $\xi_{ji} \simeq \gamma_{ji}$, E_c is the threshold energy for relocation and $R_c = AE_c^\alpha$, the minimum distance for relocation. If the restriction $|z| > R_c$ is not imposed then the relocation function would diverge due to the formulation including the effect of large numbers of small displacements, which cannot be counted as true relocated atoms.

Diffusion in the cascade. The cascade increases the mobility of atoms for as long as the energy is distributed within a small volume of the target. To account for such motion a diffusion current is introduced:

$$J_i^{(\text{diff})}(x, t) = -\frac{\partial}{\partial X}[D_i(x, t)n_i(x, t)]. \tag{5.150}$$

Here $D_i(x, t)$ is the diffusion coefficient of the i atoms. It is often obtained assuming that the 'temperature' within the collision cascade is approximated by

$$T(r) \simeq \frac{F_D(r)}{Nk},$$

where F_D is the energy density deposited by elastic collisions, k being the Boltzmann constant. Using the temperature above as the initial distribution of temperature one has first to solve the heat diffusion equation (Vineyard, 1976),

$$\frac{\partial T}{\partial t} = \frac{C}{\kappa}\nabla^2 T,$$

where C is the heat capacity per unit mass and κ is the thermal conductivity. Then, one assumes that the diffusion coefficient relates to temperature as

$$D_i(T) \simeq D_i^{(0)} \exp\left[-U_i/(kT)\right].$$

Here $D_i^{(0)}$ is the prefactor of the diffusion coefficient, which may depend on the instantaneous concentration of vacancies and intersitials within the target, and U_i is the activation energy for the diffusion of atom i. One can finally obtain the diffusion coefficient as

$$D_i(x) = 2\pi D_i^{(0)} I \int d\tau \int d\rho\, \rho\, \exp[-U_i/(kT)], \tag{5.151}$$

where ρ denotes the position measured along a plane perpendicular to the beam.

The integration upon τ extends over the cooling down of the cascade, i.e. 10^{-12} s and one assumes here that there is no overlap between cascades. In this regard, τ should not be confused with t entering equation (5.150), where the latter accounts for variations in the deposited energy F_D on a much longer time-scale. Some authors have utilized this so-called thermal spike approach as the sole relocation mechanism within the cascade (see Peak and Averback (1985), Kim *et al.* (1988) and Koponen (1992)). This appears to be well justified in cases of high-density deposited energy, such as heavy metallic targets and, particularly, for materials with a low activation energy for diffusion.

Boundary conditions. The changing composition of the surface layers means that the sputtering yields of the individual atomic components also change. A simple linear interpolation

$$Y_i = \frac{Y_i^{(0)} n_i(0, t)}{n_t(0, t)},$$

where $n_t(0, t) = \sum_i n_i(0, t)$, has been found to be inadequate (Wadsworth *et al.*, 1990). Instead it is better to use

$$Y_i = \frac{F_{Di}}{e_i},$$

where F_{Di} is the energy deposited onto the atoms i at the surface. The quantity e_i can be calculated by defining e_{ij}, the energy required to eject unit volume of i atoms from a j material matrix so that

$$e_i = \sum e_{ij} n_j.$$

The same approach to non-linear interpolation can also be used for the range functions within the mixture.

5.5.3 The diffusion approximation

The mass current in equation (5.143) can be cast into a much simpler form by using the fact that the relocation function, $F_i(x, z)$ is strongly peaked around $z = 0$. Hence, according to approximations in equations (5.146)–(5.149) we can write $F_i(x, z) = F_i(x) f_i(z)$, where the function $f_i(z)$ satisfies

$$\int_{-\infty}^{+\infty} f_i(z) \, dx = 1.$$

Equation (5.134) thus gives

$$J_i^{(cas)}(x) = I \int_0^x dx' \, n_i(x', t) F_i(x') \int_{x-x'}^{\infty} dz \, f_i(z)$$

$$- I \int_x^{\infty} dx' \, n_i(x', t) F_i(x') \int_{-\infty}^{x-x'} dz \, f_i(z). \tag{5.152}$$

As a consequence of the peaked behaviour of $f(z)$ towards small z, the main contribution to integration with respect to x' in equation (5.143) proceeds from the region $x' \simeq x$; one can therefore expand $n_i(x', t)F_i(x')$ in a power series around $x' = x$,

$$n_i(x', t)F_i(x') = \sum_{n=0} \frac{(x' - x)^n}{n!} \frac{\partial^n}{\partial x^n} [n_i(x, t)F_i(x)], \tag{5.153}$$

then

$$J_i^{(\text{cas})}(x) = I \sum_{n=0} \frac{J_i^{(n)}(x)}{n!} \frac{\partial^n}{\partial x^n} [n_i(x, t)F_i(x)], \tag{5.154}$$

where

$$\begin{aligned} J_i^{(n)} &= \int_0^x dx' \, (x' - x)^n \int_{x-x'}^\infty dz \, f_i(z) \\ &- \int_x^\infty dx' \, (x' - x)^n \int_{-\infty}^{x-x'} dx \, f_i(z). \end{aligned} \tag{5.155}$$

Replacing the lower limit by $-\infty$ in the first integral, the J_i terms are no longer functions of x. Then one readily obtains

$$J_i^{(n)} = \frac{(-1)^n}{n+1} \langle z^{n+1} \rangle_i, \tag{5.156}$$

where $\langle z^n \rangle_i$ is the nth moment of the function $f_i(z)$,

$$\langle z^n \rangle_i = \int_{-\infty}^{+\infty} dz \, z^n f_i(z).$$

Retaining only first-order terms in the expansion of (5.154) yields

$$J_i^{(\text{cas})}(x) \simeq \langle z \rangle_i n_i(x, t)F_i(x) - \frac{1}{2} I \langle z^2 \rangle_i \frac{\partial}{\partial x} [n_i(x, t)F_i(x)]. \tag{5.157}$$

This form of the cascade current, known as the *diffusion approximation*, becomes quite acceptable when relocations result mainly from a large number of small displacements.

5.5.4 A mixing example

Marker broadening. If one utilizes a target containing a thin impurity layer (*marker*) the ability of ion beams to relocate the impurity among the host material can be straightforwardly determined. The relocation of the impurity among the host material will translate into a *broadening* and *shifting* of the marker with increasing dose.

The expressions previously derived will be applied to the spreading of a thin Pt marker buried at 500 Å in Si during 300 keV Xe ion bombardment (after Paine (1982)). Here, we assume that $i = 2$ stands for Si atoms whereas $i = 3$ corresponds to Pt.

First, we introducing the volume density, i.e. $\theta_i(x,t) = v_i n_i(x,t)$. Hence there results the packing constraint

$$\sum_i \theta_i(x,t) = 1.$$

To simplify we shall assume that $\theta_1(x,t)$ is negligibly small compared with those of Pt and Si, hence giving

$$\theta_{Si}(x,t) + \theta_{Pt}(x,t) = 1.$$

Secondly we set $U = 0$ and $J_i^{(diff)} = 0$. Then, the back-current can finally be written as

$$K(x,t) = \sum_{i=2,3} v_i J_i^{(cas)}(x,t).$$

If one multiplies equation (5.140) by v_{Pt} it yields

$$\frac{\partial \theta_{Pt}}{\partial t} = -\frac{\partial}{\partial x}\left[v_{Pt} J_{Pt}^{cas} - \theta_{Pt} k(x,t)\right]. \tag{5.158}$$

This equation is not only non-linear but also strongly coupled through the back-current $k(x,t)$. Substituting the cascade current by the diffusion approximation and replacing $F_i(x)$ by $F_i(x_0)$, since F_i varies slowly with depth, it finally yields

$$v_{Pt} J_{Pt}^{cas}(x,t) \simeq I\left(v_{Pt}\theta_{Pt} - D_{Pt}\frac{\partial \theta_{Pt}}{\partial x}\right), \tag{5.159}$$

where

$$v_{Pt} = z_{Pt} F_{Pt}(x_0), \quad D_{Pt} = z_{Pt}^2 F_{Pt}(x_0). \tag{5.160}$$

Hereinafter, we shall refer to V_i and D_i as the shift and spreading coefficients, respectively, since, provided that there is no back-current, they represent the mean distance and the corresponding uncertainty in ith atom relocation per unit dose.

Since, as we shall see below, the back-current is dominated by the Si current, one can thus approximate

$$k(x,t) \approx v_{Si} J_{Si}^{(cas)}(x,t).$$

Furthermore, because the distribution of Si atoms does not change appreciably with depth

$$k(x,t) \approx I v_{Si}\theta_{Si}(x,t) \approx I v_{Si}(x_0). \tag{5.161}$$

Table 5.3. *Shift and spreading coefficients*

Approximation quantity	v_i (Å^3/ion)	D_i (10^4Å^4/ion)
Straight line Xe/Pt	17	0.36
Isotropic Si recoils/Pt (1)	0	0.02
Isotropic Si recoils/Pt (2)	0	0.13
Straight line Xe/Si	26	1.20

Therefore, we can rewrite equation (5.158) as

$$\frac{\partial \theta_{Pt}}{\partial t} = I \left(v_{Si} - v_{Pt} \right) \frac{\partial \theta_{Pt}}{\partial x} + I D_{Pt} \frac{\partial^2 \theta_{Pt}}{\partial x^2}. \tag{5.162}$$

Equation (5.162) is not only totally decoupled but also reduced to a simple diffusion equation plus a drift term.

The broadening of the Pt marker as a function of the ion dose, i.e. $\phi = It$, can be readily obtained, the result being

$$\Omega^2(\phi) = D_{Pt}(x_0)\phi. \tag{5.163}$$

Similarly, the shift of the marker as a function of ion dose gives

$$\Delta X_{Pt}(\phi) = (v_{Si} - v_{Pt})\phi. \tag{5.164}$$

In order to compare the results above with experiments, we shall calculate the shift and spreading coefficients resulting from the expressions and labelled (1) and (2) in Table 5.3. Here, however, we use only one of the ion–impurity approximations, since the straight-line and isotropic ion–target and ion–impurity relocations are nearly identical insofar as an order of magnitude estimation is concerned.

To do the calculation we first need the mean energy of the Xe ions at the marker site, i.e. $E(500$ Å$)$. We can identify such an energy with the mean energy of 300 keV Xe ions after traversing a 500 Å-thick silicon foil. This can readily be done using the TRIM simulation code (Biersack and Haggmark, 1980). The result is $E(500$ Å$) \approx 150$ keV.

Secondly, we have to calculate z_{max}, the mean projected range both for Pt and for Si in Si at the maximum energy which they can receive from a 150 keV Xe projectile, i.e. 130 and 80 keV respectively. Again, using the TRIM code we find 540 Å for Pt and 1140 Å for Si.

In the case of the straight-line ion–impurity relocation equation (5.147)

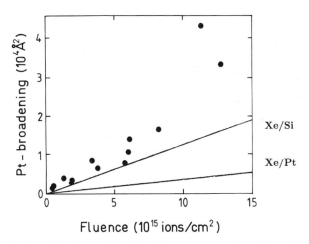

Fig. 5.20 Broadening of a Pt marker in a Si target as a function of the bombarding (300 keV) Xe-ion dose (after Paine (1982)). The straight lines are predictions from ballistic theory using the diffusion approximation.

we put $\alpha = 1$ and $m = 1/2$. For the isotropic recoil–impurity relocation equation (5.149) we have two possibilities, (1) following Sigmund and Gras-Martí (1981) we can put $\alpha = m = 0$, E_c=7.8 eV and R_c=3 Å or, (2) we may set $\alpha = 1/4$, hence $R_c^2 \propto E_c$ so that, using the Born–Mayer values for the scattering cross-section we obtain $R_c^2/E_c \approx 5.5$ Å^2 eV^{-1}. Finally, the deposited energy is obtained from TRIM simulations, the result being $F_D(500\text{Å}) \approx 100$ eV Å^{-1}. In Figure 5.20 we show the broadening for a Pt marker buried at $x_0 =500$ Å within a silicon substrate during 300 keV Xe$^+$ irradiation (Paine, 1982). Observe that the marker broadening increases linearly with increasing dose, in the same way as predicted by the collisional theory, i.e. equation (5.154). The spreading coefficient can be obtained by adding the ion/Pt relocation to one of the two recoil/Pt-relocation approximations, namely options (1) and (2) shown in Table 5.3 because the ion–impurity and recoil–impurity mechanisms are independent. As one can see in Figure 5.20, theoretical results are significantly smaller than experimental values. This has been observed before, particularly with the approximations of Sigmund and Gras-Martí, i.e. option (1). With $m = 1/4$, i.e. option (2), the broadening is slightly larger. This can be readily explained, since for $m = 1/4$ the range of Pt increases with energy faster than it does with $m = 0$. The probability of relocations to larger distances thus increases, which is reflected in equation (5.140), where, given that $\alpha = 2m$, as m increases the relocation function will be a much slower decreasing function of $|z|$.

The shift of the marker with dose has been plotted in Figure 5.21, in which the discrepancy between experiment and calculation is large. Although

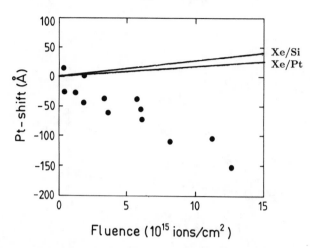

Fig. 5.21 The shift of the Pt marker in a Si target as a function of the bombarding (300 keV) Xe-ion dose (after Paine (1982)). The predictions are from ballistic theory using the diffusion approximation.

calculations predict a shift of the marker towards the surface, experimental data show just the opposite.

The results of calculations can be easily explained. The mean relocation distance for Si atoms is larger than that for Pt. Hence, there will be a net flux of Si atoms from left to right through the marker and, consequently, the marker must shift towards the surface in order to make room for the Si atoms.

The origin of the *observed* shift is not clear. The interested reader is referred to the original article by Paine. However, good agreement between experiment and theory was obtained by Wadsworth *et al.* (1990), using the more complex numerical model, for an analysis of the degradation of depth profiles during SIMS. Figure 5.22 compares their experimental and theoretical results.

The previous example has shown a number of the difficulties arising in modelling mixing. These arise because the atomic relocation is sensitive to a variety of low-energy processes, which are intrinsically non-linear and not yet completely understood. Even the electronic stopping at lower energies may have an unexpectedly large influence upon the mixing in metallic targets.

5.6 Conclusion

Boltzmann transport theory is an extremely useful tool for obtaining simple formulae that can be easily applied in many important applications. It can predict general trends quite well, for example the dependence of the

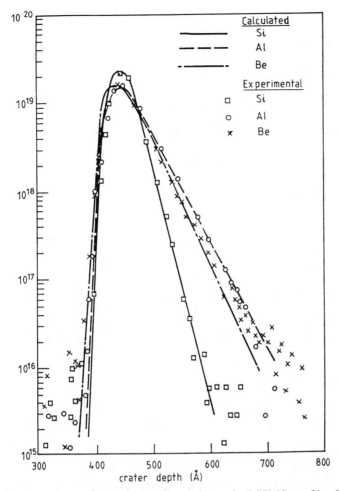

Fig. 5.22 A comparison of experimental and theoretical SIMS profiles for Si, Al and Be δ spikes in GaAs (courtesy R. Badheka). The bombarding species is O, incident at $2°$ to the surface normal with energy 2.4 keV.

sputtering yield on the energy of the bombarding species and the first-order implantation distributions. Because it is a statistical theory, it is less successful if there is target ordering (crystalline materials) or if individual mechanisms need to be explained.

In this chapter we have concentrated on obtaining approximate solutions to the linear BE. To obtain more accurate solutions is more complicated, often requiring sophisticated numerical techniques. The next chapter describes how the BE can be solved accurately, with reference to the problem of particle implantation, showing the good agreement between these accurate numerical solutions and binary collision computer simulations.

References

Abramowitz, M. and Stegun, I. Eds. (1970). *Handbook of Mathematical Functions*, Dover Publications, Inc., New York.

Ashworth, D. G., Bowyer, M. D. J. and Oven, R. (1991). *J. Phys.* D **24** 1376.

Ashworth, D .G., Bowyer, M. D. J. and Oven, R. (1995). *Nucl. Instrum. Meth.* B **100** 471.

Biersack, J. P. (1981). *Nucl. Instrum. Meth.* **182/183** 199.

Biersack, J.P. (1982). *Z. Phys.* A **305** 95.

Biersack, J. P. and Haggmark, L. G. (1980). *Nucl. Instrum. Meth.* **174**, 257.

Bowyer, M. D. J., Ashworth, D. G. and Oven, R., (1992). *J. Phys.* D **25** 1619; (1994) *Rad. Eff. Defects Solids* **130/131** 535.

Case, K. M. and Zweifel, P. F. (1967). *Linear Transport Theory*, Addison-Wesley, Reading, Massachussets.

Chou, P. S. and Ghoniem, N. M. (1986). *J. Nucl. Mater.* **141/143**, 216.

Cramér, H. (1958). *Elements of Probability Theory*, John Wiley & Sons, New York.

Furukawa, S. Matsumura, H. and Ishiwara, H., (1972). *Jap. J. Appl. Phys.* **11** 134.

Gardiner, C. W. (1985). *Handbook of Stochastic Methods for Physics, Chemistry and the Natural Sciences* (2nd edition), Springer-Verlag, Berlin.

Gibbons, J. F. and Christel, L. A. (1984). *Ion Implantation and Beam Processing*. Ed. J. S. Williams and J. M Poate, Academic Press Australia, 59.

Kim, S. J., Nicolet, M. A., Averback, R. S. and Peak, D. (1988). *Phys. Rev.* B **37** 38.

Koponen, I. (1992). *J. Appl. Phys.* **72** 1194.

Lindhard, J., Nielsen, V. and Scharff, M. (1968). *Mat. Fys. Medd. Dan. Vid. Selsk.* **36** 10.

Morse, P. M. and Feschbach, H. (1953). *Methods of Theoretical Physics*. McGraw-Hill Book Co., New York.

Paine, B. M. (1982). *J. Appl. Phys.* **53** 6826.

Peak, D. and Averback, R. S. (1985). *Nucl. Instrum. Meth.* B **178** 561.

Shohat, J. A. and Tamarkin J. D. (1943). *The Problem of Moments*, American Mathematical Society, New York.

Sigmund, P. (1969). *Phys. Rev.* **184** 383.

Sigmund, P. (1972). *Rev. Roum. Phys.* **17** 823, 969, 1079.

Sigmund, P. (1981). *Sputtering by Particle Bombardment I* Ed. R. Berhisch, Springer-Verlag, Berlin, ch 2.

Sigmund, P. and Gras-Martí, A. (1981). *Nucl. Instrum. Meth.* **182/183** 25.

Thompson, M. W. (1968). *Phil. Mag.* **18** 377.

Vineyard, G. H. (1976). *Rad. Eff. Defects Solids* **29** 245.

Wadsworth, M., Armour, D. G., Badheka, R. and Collins, R. (1990). *Int. J. Num. Modelling* **3** 157.

Williams, M. M. R. (1971). *Mathematical Methods in Particle Transport Theory*, Butterworth and Co., London.

Winterbon, K. B., Sigmund, P. and Sanders, J. B. (1970). *Mat. Fys. Medd. Dan. Vid. Selsk.* **87** 14.

Ziegler, J. F., Biersack, J. P. and Littmark, U. (1985). *The Stopping and Range of Ions in Solids*, Pergamon Press, New York.

6

The rest distribution of primary ions in amorphous targets

6.1 Introduction

In many applications it is the rest distribution of the implanted (or primary) ions that is of principal importance, e.g. in the doping of semi-conductors (Sze, 1988). We examine this in detail because of its intrinsic importance and also because we can illustrate some modern statistical and numerical techniques applied to transport theory in a little more detail than described in the previous chapter. The penetration of ions into amorphous targets is described most simply by using a statistical transport model. The use of this model has the advantage that two methods exist for the prediction of the rest distribution of ions: the solution of transport equations (TEs) and Monte Carlo (MC) simulation. A statistical model is essential to the construction of TEs and the computational efficiency that it affords MC simulation is necessary in order to obtain good statistics.

In several ways the MC and TE methods are complementary. In direct form the MC method treats an explicit sequence of collisions, so the target composition can change on arbitrary boundaries (in space and time). The rest distribution is built up from a large number of ion trajectories, the statistical precision of which depends directly on this number. Hence, the use of the MC method is dependent on the necessary statistical precision being obtained in a 'sensible' amount of CPU time. On the other hand, Lindhard-type TEs assume a target that satisfies space (and time) translational invariance. The only target to satisfy this condition is infinite and homogeneous. However, the Lindhard-type TEs are able to trade such severe constraints for an ability to produce, efficiently, the moments of the rest distribution as a function of the implant energy. The production of such functions, inside a process simulator, is necessary in order to provide an optimisation capability.

The classical Boltzmann and Lindhard TEs described in Chapter 5 assume

161

the *gas* target model in which the distribution of free flight path lengths is exponential. In order to increase the efficiency of MC codes, the *gas* model can be replaced by the *liquid* model, in which the distribution of free flight path lengths is a delta function.

In this chapter it will be shown how an algorithm for MC simulation and a vector range transport equation (of the Lindhard type) can be developed from a statistical model that includes a general distribution for the free flight path length. It will also be shown how the standard TE solution, employing spatial moments and Legendre polynomial expansions, may be modified in order to incorporate the *liquid* model. Furthermore, it will be shown that a numerical solution technique (Winterbon, 1986) originally derived for the *gas* model, is also suitable for the *liquid* model. Once complete, the moments solution still leaves the problem of profile construction. This is problematical because a unique profile cannot be constructed from a finite set of moments. If the set becomes too large, then the numerical TE solution is slow. However, it has been shown (Hofker *et al.*, 1975) that realistic one-dimensional profiles can be obtained from the sets including fourth-order moments. The Pearson system of frequency curves, which is the most popular choice for curve construction, is examined in detail. A practical method for constructing two-dimensional profiles, which has evolved since the mid 1980s, is reviewed. Finally, the results from TE and MC computer codes are compared.

6.2 Ion–solid interaction

As an ion penetrates a solid it undergoes a sequence of collisions with the target atoms until it comes to rest. A simplified model of this interaction is a sequence of instantaneous binary nuclear collisions separated by free flight path lengths (straight line segments) over which the ion experiences a continuous (non-local) electronic energy loss. The collisions are separated, i.e. the state of an ion after a collision depends solely on the state of the ion before the collision. So, in order to construct an algorithm for MC simulation, or to set up a transport equation (TE), it is simply necessary to characterise a single binary collision together with the free flight path length that occurs either before or after the collision.

Figure 6.1 depicts a simplified encounter between an energetic ion with atomic number Z_1 and mass m_1, and a stationary target atom with atomic number Z_2 and mass m_2. The state of the ion before the collision is $[E, \hat{u}, r]$, where E is the pre-flight energy, $\hat{u}$ is a unit vector in the pre-flight velocity direction and r is a vector that specifies the pre-flight location. In flight, the ion experiences a non-local electronic energy loss, T_l. The state of the

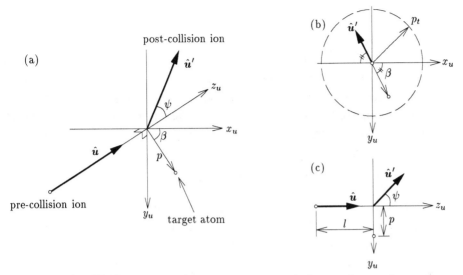

Fig. 6.1 A simplified encounter between an energetic ion and a stationary target atom: (a), three-dimensional representation; (b), projection onto the x_u–y_u plane; (c), the plane containing the target atom and the unit vectors. The symbols are defined in the text.

ion after the flight is $[E_l, \hat{u}, r']$, where E_l is the post-flight energy given by $E_l = E - T_l$, the velocity direction is unchanged and r' is a vector that specifies the post-flight location given by $r' = r + \hat{u}l$. If the path length, l, is specified, then T_l is determined from

$$l = -\frac{1}{N} \int_E^{E-T_l} \frac{\mathrm{d}E}{S_e(E)},$$ (6.1)

where N is the number of target atoms per unit volume and $S_e(E)$ is the electronic stopping power.

In practice T_l is computed by dividing the path length l into n short segments, within which $S_e(E)$ is assumed constant. If the ith segment has length l_i, then the ion energy after segment i is given by the recurrence relation $E_{i+1} = E_i - NS_e(E_i)l_i$, where $E_0 = E$. In this case, T_l is given by $E_0 - E_n$. Strictly speaking, the function $S_e(E)$ is determined by one variable and three parameters,

$$S_e = S_e(E; Z_1, Z_2, m_1)$$ (6.2)

and, therefore, T_l is determined by two variables and four parameters,

$$T_l = T_l(E, l; Z_1, Z_2, m_1, N).$$ (6.3)

The functional forms of equations that determine $S_e(E)$ are given in Chapter 4.

The post-flight state of the ion after the collision is $[E', \hat{u}', r']$, where E' is the post-collision energy and $\hat{u}'$ is a unit vector in the post-collision velocity direction. The collision depicted is instantaneous but, in reality, the interaction is finite in space and time. The reality is embraced in the framework of the binary collision approximation (see Chapter 2), in which the vectors $\hat{u}$ and $\hat{u}'$ define asymptotes to the real path of the ion. The target atom is located in a plane normal to $\hat{u}$. In this plane, the location of the target atom is specified by the impact parameter, p, and the azimuthal scattering angle, β.

The interaction is governed by a conservative central force so, once the target atom has been selected, the movement of the particles is restricted to a plane. (The specified plane contains $\hat{u}$ and the target atom.) In this case, the interaction with the target atom results in the laboratory scattering angle, ψ, and the nuclear energy transfer, T_n, being independent of β. The equations that determine the quantities ψ and T_n are derived from classical scattering theory (see Chapter 2). In this theory, the variables E and p and the parameters Z_1, Z_2, m_1 and m_2 are used to compute $\bar{\theta}$, the scattering angle in the centre-of-mass coordinate system, i.e. $\bar{\theta}$ is a function of two variables and four parameters:

$$\bar{\theta} = \bar{\theta}(E, p; Z_1, Z_2, m_1, m_2). \tag{6.4}$$

The quantities ψ and T_n are obtained independently from $\bar{\theta}$ by using

$$\psi = \tan^{-1}\left(\frac{m_2 \sin \bar{\theta}}{m_1 + m_2 \cos \bar{\theta}}\right) \equiv \psi(E, p; Z_1, Z_2, m_1, m_2), \tag{6.5}$$

$$T_n = \frac{4m_1 m_2}{(m_1 + m_2)^2} E \sin^2\left(\frac{\bar{\theta}}{2}\right) \equiv T_n(E, p; Z_1, Z_2, m_1, m_2). \tag{6.6}$$

Once the values of ψ and T_n are known, the post-collision state, $[E', \hat{u}', r']$, is determined from the post-flight/pre-collision state. The energy E' is simply

$$E' = E_l - T_n, \tag{6.7}$$

where $T_n = T_n(E_l, p)$ and the velocity direction $\hat{u}'$ is given by a function of three variables,

$$\hat{u}' = \hat{u}'(\hat{u}, \psi, \beta), \tag{6.8}$$

where $\psi = \psi(E_l, p)$. In actual calculations it is necessary to follow the motion with respect to a fixed (or reference) coordinate system. To this end, the following convention is adopted. *The ion is implanted along the z-axis of a Cartesian coordinate system and enters the target at z = 0. The target surface (of infinite extent) occupies the x–y plane.* Using this convention, $\hat{u}$ is defined by three direction cosines, (η_x, η_y, η_z), with respect to the axes x, y and z. The direction cosines, $(\eta_x', \eta_y', \eta_z')$, corresponding to $\hat{u}'$ can be obtained from the product of four simple rotation matrices and a unit vector and are given by

$$
\begin{bmatrix} \eta_x' \\ \eta_y' \\ \eta_z' \end{bmatrix} = R_{x-y}(\alpha) R_{z-x}(\omega) R_{x-y}(\beta) R_{z-x}(\psi) \begin{bmatrix} 0 \\ 0 \\ 1 \end{bmatrix}, \tag{6.9}
$$

where the angles ω and α are defined in Figure 6.2 and

$$
R_{z-x}(v) = \begin{bmatrix} \cos v & 0 & \sin v \\ 0 & 1 & 0 \\ -\sin v & 0 & \cos v \end{bmatrix}, \quad R_{x-y}(v) = \begin{bmatrix} \cos v & -\sin v & 0 \\ \sin v & \cos v & 0 \\ 0 & 0 & 1 \end{bmatrix}. \tag{6.10}
$$

Finally, the substitution of

$$
\cos \omega = \eta_z, \quad \sin \omega = (1 - \eta_z^2)^{1/2}, \quad \cos \alpha = \eta_x/(1 - \eta_z^2)^{1/2},
$$
$$
\sin \alpha = \eta_y/(1 - \eta_z^2)^{1/2} \tag{6.11}
$$

into equation (6.9) gives

$$
\eta_x' = \eta_x \cos \psi + \frac{\sin \psi}{(1 - \eta_z^2)^{1/2}} (\eta_x \eta_z \cos \beta - \eta_y \sin \beta), \tag{6.12}
$$

$$
\eta_y' = \eta_y \cos \psi + \frac{\sin \psi}{(1 - \eta_z^2)^{1/2}} (\eta_y \eta_z \cos \beta + \eta_x \sin \beta), \tag{6.13}
$$

$$
\eta_z' = \eta_z \cos \psi - (1 - \eta_z^2)^{1/2} \sin \psi \cos \beta. \tag{6.14}
$$

N.B. The use of equation (6.9) assumes that β is a uniform random variable (between 0 and 2π) because the azimuthal orientation of the axis x_u–y_u, shown in Figure 6.1, is pre-determined, before each scattering event, by the angle α.

6.3 A random solid

The statistical transport model assumes that the arrangement of the target atoms is totally randomised after each collision, i.e. the target is structureless and memoryless. As a result, a sequence of collisions is described by randomly selecting the location of the next collision partner relative to the pre-flight location and velocity direction of the ion. From Section 6.2, the

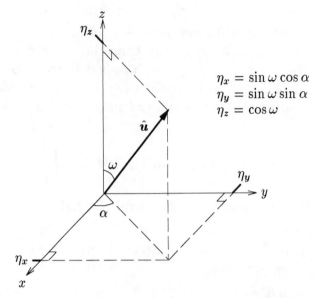

$$\eta_x = \sin \omega \cos \alpha$$
$$\eta_y = \sin \omega \sin \alpha$$
$$\eta_z = \cos \omega$$

Fig. 6.2 A unit vector in the pre-collision velocity direction defined in terms of a polar angle ω and an azimuthal angle α.

location of a collision partner relative to the pre-flight ion is determined by the variables l, p and β. Thus, to select randomly the location of the next collision partner the variables l, p and β become independent random variables with probability density functions: $f_1(l)$, $f_2(p)$ and $f_3(\beta)$.

These density functions are defined as follows: $f_1(l)\,dl$ is the probability that, following the free flight path length l, the ion undergoes a scattering event in the interval between l and $l + dl$; $f_2(p)\,dp$ is the probability of an impact parameter between p and $p + dp$ and $f_3(\beta)\,d\beta$ is the probability of an azimuthal scattering angle between β and $\beta + d\beta$. They are normalised such that

$$\int_0^\infty f_1(l)\,dl = 1, \quad \int_0^\infty f_2(p)\,dp = 1, \quad \int_0^{2\pi} f_3(\beta)\,d\beta = 1. \tag{6.15}$$

Referring back to Section 6.2, the ion interacts with a target atom that is located in a plane normal to $\hat{u}$. The location of the target atom is constrained by a circle of radius p_t, centred on the point of deflection, i.e. $0 \le p < p_t$. If a scattering event occurs, then the target atom lies somewhere inside this circle. All locations inside the circle are equally likely because the arrangement of the target atoms is truly random. In this case, the probability density functions $f_2(p)$ and $f_3(\beta)$ are given by

$$f_2(p) = \frac{2\pi p}{\pi p_t^2}\Theta(p_t - p), \quad f_3(\beta) = \frac{1}{2\pi}, \tag{6.16}$$

where Θ is the Heaviside step function. The same equations arise from a homogeneous flux of ions incident on a single scattering centre with a differential nuclear scattering cross-section, $d\sigma_n = 2\pi p\, dp$, and a (finite) total nuclear scattering cross-section, $\sigma_{nt} = \pi p_t^2$.

The most commonly used form of $f_1(l)$ is called the *gas* model, which likens the target to a dense gas. A simple analysis, based on Poisson statistics, results in the well-known exponential distribution

$$f_1(l) = N\sigma_{nt} \exp(-N\sigma_{nt}l). \tag{6.17}$$

The most important measure of $f_1(l)$ is the mean free flight path length, λ, defined as

$$\lambda = \int_0^\infty l f_1(l)\, dl. \tag{6.18}$$

Substituting equation (6.17) into equation (6.18) gives

$$\lambda = \frac{1}{N\sigma_{nt}}. \tag{6.19}$$

The geometric interpretation of equation (6.19) is of scattering centres individually located inside cylindrical volumes of mean length λ and base area σ_{nt}, where the product $\sigma_{nt}\lambda$ is fixed equal to N^{-1}. Equation (6.19) establishes an important relationship between λ and σ_{nt} and, in order to preserve the atomic density of the target, it is essential to satisfy this relationship. For example, if λ is fixed equal to the approximate mean interatomic separation, $N^{-1/3}$, then $p_t = \pi^{-1/2}N^{-1/3}$.

In the *gas* model, the free flight path length can take extreme values: $l \gg \lambda$ and $l \ll \lambda$. It has been argued (Biersack, 1982, Eckstein, 1991) that extreme values for l are not representative of an amorphous solid. A simple alternative to the *gas* target model is called the *liquid* target model (Biersack, 1982, Eckstein, 1991, Miyagawa and Miyagawa, 1983). In this model, l is fixed equal to λ and $f_1(l)$ is the delta function given by

$$f_1(l) = \delta(l - \lambda). \tag{6.20}$$

The density functions for $f_2(p)$ and $f_3(\beta)$ are unchanged from the *gas* model, i.e. the distribution of target atoms in the plane normal to $\hat{u}$ is uniform. Clearly, the *liquid* model uses two random variables to select the next collision partner, whereas the *gas* model uses three. The need for one less random variable improves the efficiency of MC codes and the use of two random variables (degrees of freedom), instead of three, imposes a sense of local order on the target structure.

6.4 MC simulation

The MC simulation of ion implantation follows the trajectory of an ion that collides with a sequence of randomly selected collision partners. The ion starts with an energy, E, equal to the implantation energy, E_0; a velocity direction, $\hat{u}$, equal to the beam direction, $\hat{u}_0$, and a location, $r = (r_x, r_y, r_z)$, at the origin, $(0,0,0)$. The trajectory is followed until the ion energy falls below an energy, E_s, at which the ion is assumed to stop. The rest distribution is obtained by recording the termination coordinates of a large number of trajectories.

6.4.1 Non-uniform random variables

From Section 6.3, the location of a collision partner is determined by three random variables l, p and β. The probability density function $f_3(\beta)$ is uniform, but the functions $f_1(l)$ and $f_2(p)$ are non-uniform. In the majority of computer simulations, non-uniform random variables are obtained from uniform random variables because high-speed algorithms are widely available that produce uniform pseudo-random numbers (Knuth, 1981, Marsaglia, 1985). Two general methods that can be used to transform uniform random numbers to non-uniform random numbers are the *inversion* and *rejection* methods (Duderstadt and Martin, 1979, Morgan, 1984). The general form of the rejection method is too slow to be of interest. (It involves repeated evaluation of the probability density function using two random numbers per evaluation.) The inversion method is equivalent to sampling the cumulative distribution function of the random variable. The cumulative distribution functions that correspond to l, p and β are given by

$$C_1(l) = \int_0^l f_1(l')\,\mathrm{d}l', \quad C_2(p) = \int^p f_2(p')\,\mathrm{d}p, \quad C_3(\beta) = \int_0^\beta f_3(\beta')\,\mathrm{d}\beta'. \quad (6.21)$$

Such functions are monotonically increasing and range from 0 to 1 as the variable ranges from 0 to ∞.[1] Thus, for each value of the variable there is a unique value of the function. More importantly, the relationship between the probability density function and the cumulative distribution function ensures that the function values $C_1(l)$, $C_2(p)$ and $C_3(\beta)$ are distributed uniformly between 0 and 1, c.f. the derivation in Duderstadt and Martin (1979), pp 536–537. For these reasons, the cumulative distribution function can be sampled uniformly, i.e. unique values of l, p and β can be found that

---

[1] β ranges from 0 to 2π.

satisfy

$$C_1(l) = R_1, \quad C_2(p) = R_2, \quad C_3(\beta) = R_3, \tag{6.22}$$

where R_1, R_2 and R_3 are (independent) uniform random numbers between 0 and 1. If the inverse cumulative distribution functions exist, then sampling is equivalent to computing the inverse functions:

$$l = C_1^{-1}(R_1), \quad p = C_2^{-1}(R_2), \quad \beta = C_3^{-1}(R_3). \tag{6.23}$$

The results for the *gas* model are

$$l = -\lambda \ln(1 - R_1), \quad p = p_t \sqrt{R_2}, \quad \beta = 2\pi R_3. \tag{6.24}$$

(The equation for l is simplified by noting that the distributions of R_1 and $1 - R_1$ are identical.) The only change for the *liquid* model is that $l = \lambda$.

6.4.2 Algorithms

The necessary background is now in place to construct an algorithm for MC simulation. An algorithm for the *gas* model, in pseudo-code, is

for $ion = 1$ **to** *number_of_ions*
$\quad E = E_0, \quad \hat{u} = \hat{u}_0, \quad r = (0,0,0)$
$\quad$ **repeat**
$\qquad\quad$ generate R_1, R_2 and R_3
$\qquad\quad l = C_1^{-1}(R_1)$
$\qquad\quad T_l = T_l(E, l)$
$\qquad\quad E = E - T_l$
$\qquad\quad p = C_2^{-1}(R_2), \quad \beta = C_3^{-1}(R_3)$
$\qquad\quad T_n = T_n(E, p), \quad \psi = \psi(E, p)$
$\qquad\quad \hat{u}' = \hat{u}'(\hat{u}, \psi, \beta)$
$\qquad\quad E = E - T_n$
$\qquad\quad r = r + \hat{u}l$
$\qquad\quad \hat{u} = \hat{u}'$
$\quad$ **until** $(E < E_s)$ **or** $(r_z < 0)$
$\quad$ *update_product_sum*(r_x, r_z)
$\quad$ *update_histogram*(r_x, r_z)
next *ion*

For clarity, the dependences on N, Z_1, Z_2, m_1 and m_2 of the functions T_l, T_n and ψ have been omitted. The changes necessary for the *liquid* model are to remove the random number R_1 and to replace the statement $l = C_1^{-1}(R_1)$ with $l = \lambda$.

In practice, the rest distribution takes the form of a histogram. The profile in two dimensions is the histogram $H_{j,k}$ where j and k are the bin numbers on the x and z axes respectively. The procedure *update_histogram* takes the form

update_histogram(x, z)

> **begin**
>
> $$j = \text{int}(x/\Delta x) + 1, \ k = \text{int}(z/\Delta z) + 1$$
> $$H_{j,k} = H_{j,k} + 1$$
>
> **end**

where x is the stopping co-ordinate on the x axis, z is the stopping co-ordinate on the z axis, Δx is the bin width on the x axis, Δz is the bin width on the z axis and int is a function returning the integer part of the argument. The vertical profile in one dimension (the projected range distribution) is the histogram H_k given by

$$H_k = \sum_j H_{j,k}. \tag{6.25}$$

The spatial moments of the rest distribution can be calculated from the histograms. For example, the mean projected range, $\langle z \rangle$, is given by

$$\langle z \rangle = \frac{\sum_k (k\,\Delta z) H_k\,\Delta z}{\sum_k H_k\,\Delta z}, \tag{6.26}$$

but moments calculated in this way include quantisation error due to the finite bin size. It is a simple matter to avoid histograms by maintaining product sums of the rest co-ordinates. To compute the first three spatial moments the procedure *update_product_sums* takes the form

update_product_sum(x, z)

> **begin**
>
> $$sum\_z = sum\_z + z$$
> $$sum\_z^2 = sum\_z^2 + z^2$$
> $$sum\_x^2 = sum\_x^2 + x^2$$
>
> **end**

and once the trajectory simulation is finished the moments $\langle z \rangle$, $\langle z^2 \rangle$ and $\langle x^2 \rangle$ are given by

$$\langle z \rangle = \frac{sum\_z}{number\_of\_ions}, \ \langle z^2 \rangle = \frac{sum\_z^2}{number\_of\_ions},$$
$$\langle x^2 \rangle = \frac{sum\_x^2}{number\_of\_ions}. \tag{6.27}$$

6.5 Transport equations

Transport equations (TEs) of the Lindhard type describe the probability density function of the ion rest position. The simplest way to construct a TE is to consider the first scattering event that occurs when the ion enters the target. The geometry associated with the first scattering event is shown in Figure 6.3(a). The following derivation (Bowyer, Ashworth and Oven, 1994) employs features of the original LSS derivation (Lindhard, Scharff and Schiøtt, 1963) and of its extension by Brice (1971).

6.5.1 Derivation of a Lindhard-type TE

Let $F(E, \hat{u}, r; E_s)\,dr$ represent the probability that an ion starting at the origin, with initial energy E and velocity direction $\hat{u}$, reaches the cut-off energy E_s and comes to rest in the element of volume dr centred on the vector range r. $F(E, \hat{u}, r; E_s)$ is normalised such that

$$\int_r F(E, \hat{u}, r; E_s)\,dr = 1, \tag{6.28}$$

where the integration is performed over all r. If E is less than E_s then the ion does not travel any distance, hence (Brice, 1971)

$$F(E, \hat{u}, r; E_s) = \delta(r), \tag{6.29}$$

where $\delta(r)$ is the three-dimensional Dirac delta function. If E is greater than E_s, then the probability that the ion travels a free flight path length between l and $l + dl$ and then undergoes a scattering event with impact parameter between p and $p + dp$ and azimuthal scattering angle between β and $\beta + d\beta$ is

$$f_1(l)\,dl\ f_2(p)\,dp\ f_3(\beta)\,d\beta. \tag{6.30}$$

The state of the ion after the first scattering event is $[E - T_l - T_n, \hat{u}', \hat{u}l]$, where $T_l = T_l(E, l)$, $T_n = T_n(E - T_l, p)$, $\psi = \psi(E - T_l, p)$ and $\hat{u}' = \hat{u}'(\hat{u}, \psi, \beta)$. If the ion undergoes a scattering event at $\hat{u}l$, then it has a probability density $F(E', \hat{u}', r'; E_s)$ of coming to rest at r, where $E' = E - T_l - T_n$ and $r' = r - \hat{u}l$. The essential feature of the LSS derivation is the form of this probability. From Figure 6.3(b), the arguments of E' and $\hat{u}'$ are the initial conditions for the ion with a new origin $(\eta_x l, \eta_y l, \eta_z l)$ and r' is the additional distance (measured from the new origin) necessary to reach r (measured from the old origin). Such action is possible with a target that satisfies space (and time) translational invariance.

In this case, the surface at $z = 0$ is merely a reference plane and interactions

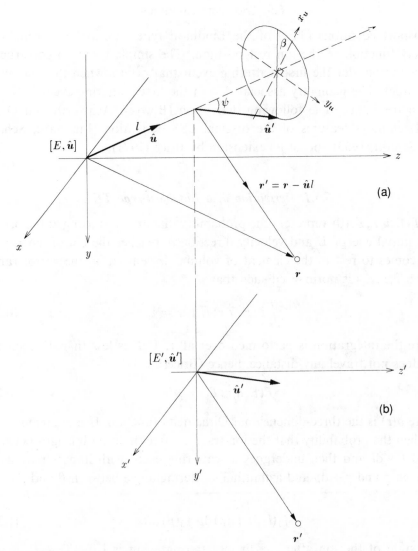

Fig. 6.3 The geometry of the first scattering event: (a) the pre-flight and post-collision states of the ion in the reference co-ordinate system; (b) the post-collision state of the ion in a new co-ordinate system, offset by $\hat{u}l$ from the reference co-ordinate system.

can occur inside the negative half plane ($z < 0$), i.e. outside the surface of a real target. Ions that obtain negative depths while slowing down and then eventually come to rest in the positive half plane ($z \geq 0$) can lead to erroneous results. Fedder and Littmark (1981) presented a treatment of real targets that employs a Lindhard-type TE.

From the above paragraph, the probability density that the ion travels a free flight path length between l and $l + dl$, gets scattered into the state $[E', \hat{u}', \hat{u}l]$ and eventually comes to rest at r, is

$$f_1(l) \, dl \, f_2(p) \, dp \, f_3(\beta) \, d\beta \, F(E', \hat{u}', r - \hat{u}l; E_s). \tag{6.31}$$

Summing equation (6.31) over all possible states with $E' \geq E_s$ gives the probability density that the ion will scatter and come to rest at r:

$$\int_0^{2\pi} \int_0^\infty \int_0^\infty f_1(l) f_2(p) f_3(\beta) F(E', \hat{u}', r - \hat{u}l; E_s) \Theta(E' - E_s) \, dl \, dp \, d\beta. \tag{6.32}$$

If either the electronic energy loss or the scattering event causes E' to fall below E_s then the ion has stopped at $r = \hat{u}l$. Hence, by equation (6.29), the probability density of an ion stopping in the first collision is

$$\int_0^\infty \int_0^\infty f_1(l) f_2(p) \delta(r - \hat{u}l) \Theta(E_s - E') \, dl \, dp. \tag{6.33}$$

The sum of the probability densities in equations (6.32) and (6.33) gives the original probability density

$$F(E, \hat{u}, r; E_s) = \int_0^{2\pi} \int_0^\infty \int_0^\infty f_1(l) f_2(p) f_3(\beta) F(E', \hat{u}', r - \hat{u}l; E_s)$$
$$\times \Theta(E' - E_s) \, dl \, dp \, d\beta \tag{6.34}$$
$$+ \int_0^\infty \int_0^\infty f_1(l) f_2(p) \delta(r - \hat{u}l) \Theta(E_s - E') \, dl \, dp.$$

This is an integral equation (IE) governing the original probability density function.

6.5.2 *Moments solution of a plane source TE*

In order to simplify development, a plane source density function is introduced, where $F(E, \eta, z; E_s) \, dz$ is defined as the probability that an ion of initial energy E reaches the cut-off energy E_s at a depth between z and $z + dz$ when the ion starts in the direction $\cos^{-1} \eta$ with respect to the z axis. This distribution is normalised such that

$$\int_{-\infty}^\infty F(E, \eta, z; E_s) \, dz = 1. \tag{6.35}$$

The point source density function $F(E, \hat{u}, r; E_s)$ can be written as $F(E, \eta, \alpha, x, y, z; E_s)$, where α is the azimuthal angle defined in Figure 6.2. The plane source density function is then related to the point source density function by

$$F(E, \eta, z; E_s) = \frac{1}{2\pi} \int_0^{2\pi} \int_{-\infty}^\infty \int_{-\infty}^\infty F(E, \eta, \alpha, x, y, z; E_s) \, dx \, dy \, d\alpha. \tag{6.36}$$

Performing these integrations on both sides of equation (6.34), and using the fact that

$$\frac{1}{2\pi} \int_0^\infty \int_0^{2\pi} \int_{-\infty}^\infty \int_{-\infty}^\infty g(l)\delta(\mathbf{r} - \hat{\mathbf{u}}l)\,\mathrm{d}x\,\mathrm{d}y\,\mathrm{d}\alpha\,\mathrm{d}l = \int_0^\infty g(l)\delta(z - \eta l)\,\mathrm{d}l, \quad (6.37)$$

where $g(l)$ is an arbitrary function of l, gives a TE for the plane source density function:

$$
\begin{aligned}
F(E,\eta,z;E_s) = {} & \int_0^{2\pi} \int_0^\infty \int_0^\infty f_1(l)f_2(p)f_3(\beta)F(E',\eta',z - \eta l;E_s) \\
& \times \Theta(E' - E_s)\,\mathrm{d}l\,\mathrm{d}p\,\mathrm{d}\beta \\
& + \int_0^\infty \int_0^\infty f_1(l)f_2(p)\delta(z - \eta l)\Theta(E_s - E')\,\mathrm{d}l\,\mathrm{d}p.
\end{aligned}
\tag{6.38}
$$

The dependence of F on depth can be removed by taking spatial moments of equation (6.38) and the dependence on velocity direction can be decoupled by expanding in terms of Legendre polynomials, $P_k(\eta)$, such that

$$
\begin{aligned}
F^n(E,\eta;E_s) &= \int_{-\infty}^\infty F(E,\eta,z;E_s)z^n\,\mathrm{d}z \\
&\equiv \sum_{k=0}^n (2k + 1)F_k^n(E;E_s)P_k(\eta).
\end{aligned}
\tag{6.39}
$$

Next, let $z_1 = z - \eta l$, followed by a binomial expansion of $(z_1 + \eta l)^n$. The integration over the azimuthal angle is carried out by utilising the addition theorem for Legendre polynomials, resulting in

$$
\begin{aligned}
\sum_{j=0}^n (2j + 1)F_j^n(E;E_s)P_j(\eta) = {} & \int_0^\infty \int_0^\infty f_1(l)f_2(p)\sum_{s=0}^n {}^nC_s(\eta l)^{n-s} \\
& \times \sum_{k=0}^s (2k + 1)F_k^s(E';E_s)P_k(\eta)P_k(\cos\psi)\Theta(E' - E_s)\,\mathrm{d}l\,\mathrm{d}p \\
& + \int_0^\infty \int_0^\infty f_1(l)f_2(p)(\eta l)^n\Theta(E_s - E')\,\mathrm{d}l\,\mathrm{d}p.
\end{aligned}
\tag{6.40}
$$

Then, multiplying both sides of equation (6.40) by $P_m(\eta)$ and integrating over all η gives

$$\sum_{j=0}^{n} (2j+1) F_j^n(E;E_s) \int_{-1}^{1} P_j(\eta) P_m(\eta) \, d\eta$$

$$= \int_{0}^{\infty} \int_{0}^{\infty} f_1(l) f_2(p) \sum_{s=0}^{n} {}^n C_s l^{n-s} \sum_{k=0}^{s} (2k+1) \qquad (6.41)$$

$$\times F_k^s(E';E_s) \int_{-1}^{1} \eta^{n-s} P_k(\eta) P_m(\eta) \, d\eta \, P_k(\cos \psi) \Theta(E'-E_s) \, dl \, dp$$

$$+ \int_{-1}^{1} \eta^n P_m(\eta) \, d\eta \int_{0}^{\infty} \int_{0}^{\infty} f_1(l) f_2(p) l^n \Theta(E_s - E') \, dl \, dp.$$

Finally, after using the orthogonality relation for Legendre polynomials (Spiegel, 1980)

$$\int_{-1}^{1} P_j(\eta) P_m(\eta) \, d\eta = \begin{cases} 0 & m \neq j \\ \dfrac{2}{2m+1} & m = j \end{cases} \qquad (6.42)$$

to remove the summation on the left-hand side of equation (6.41), an equation governing the Legendre coefficients $F_m^n(E;E_s)$ is obtained:

$$F_m^n(E;E_s) = \frac{1}{2} \int_{0}^{\infty} \int_{0}^{\infty} f_1(l) f_2(p) \sum_{s=0}^{n} {}^n C_s l^{n-s} \sum_{k=0}^{s} (2k+1)$$

$$\times \int_{-1}^{1} \eta^{n-s} P_m(\eta) P_k(\eta) \, d\eta \, F_k^s(E';E_s) P_k(\cos \psi) \Theta(E'-E_s) \, dl \, dp \qquad (6.43)$$

$$+ \frac{1}{2} \int_{-1}^{1} \eta^n P_m(\eta) \, d\eta \int_{0}^{\infty} \int_{0}^{\infty} f_1(l) f_2(p) l^n \Theta(E_s - E') \, dl \, dp.$$

Extracting the term $s = n$ from the first summation of equation (6.43), moving it to the left-hand side and introducing $r = n - s$, results in the equation

$$\int_{0}^{\infty} \int_{0}^{\infty} f_1(l) f_2(p) [F_m^n(E;E_s) - F_m^n(E';E_s) P_m(\cos \psi) \Theta(E'-E_s)] \, dl \, dp$$

$$= \frac{1}{2} \int_{0}^{\infty} \int_{0}^{\infty} f_1(l) f_2(p) \sum_{r=1}^{n-1} {}^n C_r l^r \sum_{k=0}^{n-r} (2k+1) \int_{-1}^{1} \eta^r P_m(\eta) P_k(\eta) \, d\eta \qquad (6.44)$$

$$\times F_k^{n-r}(E';E_s) P_k(\cos \psi) \Theta(E'-E_s) \, dl \, dp$$

$$+ \frac{1}{2} \int_{-1}^{1} \eta^n P_m(\eta) \, d\eta \int_{0}^{\infty} \int_{0}^{\infty} f_1(l) f_2(p) l^n \, dl \, dp.$$

An integral *driver* term of the form derived in Chapter 5 can be seen embedded in the left-hand side of equation (6.44) and on the right-hand side there is a *source* term depending on coefficients of maximum order F_m^{n-1}.

The form of this equation is important for two reasons. Firstly, the system of integral equations constructed retains the property that the coupling between equations is one way, i.e. equations in the set for higher order moments depend only on equations for lower order moments. Secondly, for numerical solution the integral driver can be approximated by a first-order ODE and used within an iterative solution scheme (Winterbon, 1986).

6.5.3 *TEs using the gas and liquid target models*

So far the derivation has retained $f_1(l)$ and $f_2(p)$ in a general manner. On substituting equations (6.16) and (6.17) into equation (6.44), the following generator equation for the *gas* model is obtained:[1]

$$
\begin{aligned}
N \int_0^{p_t} [F_m^n(E;E_s) &- F_m^n(E - T_n;E_s)P_m(\cos\psi)\Theta(E - T_n - E_s) \\
&- \Theta(E_s - E + T_n)\delta_{n0}]2\pi p\,dp + NS_e(E)\frac{dF_m^n(E;E_s)}{dE} \\
&= \frac{n[(m+1)F_{m+1}^{n-1}(E;E_s) + mF_{m-1}^{n-1}(E;E_s)]}{2m+1},
\end{aligned}
\tag{6.45}
$$

where δ_{n0} is a Kronecker delta. The normalisation and initial conditions on F imply that $F_0^0(E;E_s) = 1$. Equation (6.45) has been derived previously, but by a different route, by Littmark and Gras-Martí (1978).

Substituting equations (6.16) and (6.20) into equation (6.44), the following generator equation for the *liquid* model is obtained (Oven, Bowyer and Ashworth 1993):

$$
\begin{aligned}
N \int_0^{p_t} &\left[F_m^n(E;E_s) - F_m^n(E';E_s)P_m(\cos\psi)\Theta(E' - E_s)\right] 2\pi p\,dp \\
&= \frac{1}{2}\lambda^{n-1}\int_{-1}^1 \eta^n P_m(\eta)\,d\eta \\
&+ \frac{1}{2}\sum_{r=1}^{n-1}{}^nC_r\lambda^r\sum_{k=0}^{n-r}(2k+1)\int_{-1}^1 \eta^r P_m(\eta)P_k(\eta)\,d\eta \\
&\times N \int_0^{p_t} F_k^{n-r}(E';E_s)P_k(\cos\psi)\Theta(E' - E_s)2\pi p\,dp.
\end{aligned}
\tag{6.46}
$$

Eight equations are necessary in order to obtain fourth-order moments. Each equation can be stated in the operator notation

$$
L_m F_m^n(E;E_s) = S_m^n(E;E_s).
\tag{6.47}
$$

---

[1] For the details of this substitution see the appendix to Oven, Bowyer and Ashworth (1993).

The *gas* model has integro-differential driver terms given by

$$L_m F_m^n(E; E_s) = N \int_0^{p_t} [F_m^n(E; E_s)$$
$$- F_m^n(E - T_n; E_s) P_m(\cos\psi)\Theta(E - T_n - E_s)] 2\pi p \, dp \qquad (6.48)$$
$$+ N S_e(E) \frac{dF_m^n(E; E_s)}{dE}$$

and source terms given by

$$S_1^1 = \frac{1}{3},$$
$$S_0^2 = 2F_1^1, \quad S_2^2 = \frac{4}{5}F_1^1,$$
$$S_1^3 = F_0^2 + 2F_2^2, \quad S_3^3 = \frac{9}{7}F_2^2, \qquad (6.49)$$
$$S_0^4 = 4F_1^3, \quad S_2^4 = \frac{8}{5}F_1^3 + \frac{12}{5}F_3^3, \quad S_4^4 = \frac{16}{9}F_3^3,$$

where the energy dependence of the coefficients $F_m^n(E; E_s)$ has been omitted. The *liquid* model has integral driver terms given by

$$L_m F_n^n(E; E_s) = N \int_0^{p_t} [F_m^n(E; E_s)$$
$$- F_m^n(E'; E_s) P_m(\cos\psi)\Theta(E' - E_s)] 2\pi p \, dp \qquad (6.50)$$

and source terms given by

$$S_1^1 = \frac{1}{3},$$
$$S_0^2 = -\frac{1}{3}\lambda + \frac{2}{3}\langle z \rangle, \quad S_2^2 = -\frac{2}{15}\lambda + \frac{4}{15}\langle z \rangle,$$
$$S_1^3 = \frac{1}{5}\lambda^2 - \frac{3}{5}\lambda\langle z \rangle + F_0^2 + 2F_2^2, \quad S_3^3 = \frac{2}{35}\lambda^2 - \frac{6}{35}\lambda\langle z \rangle + \frac{9}{7}F_2^2,$$
$$S_0^4 = -\frac{1}{5}\lambda^3 + \frac{4}{5}\lambda^2\langle z \rangle - 2(F_0^2 + 2F_2^2)\lambda + 4F_1^3, \qquad (6.51)$$
$$S_2^4 = -\frac{4}{35}\lambda^3 + \frac{16}{35}\lambda^2\langle z \rangle - \left(\frac{4}{5}F_0^2 + \frac{22}{7}F_2^2\right)\lambda + \frac{8}{5}F_1^3 + \frac{12}{5}F_3^3,$$
$$S_4^4 = -\frac{8}{315}\lambda^3 + \frac{32}{315}\lambda^2\langle z \rangle - \frac{8}{7}F_2^2\lambda + \frac{16}{9}F_3^3,$$

where the mean projected range $\langle z \rangle = 3F_1^1$ has been introduced and the energy-dependence on the coefficients $F_m^n(E; E_s)$ has been omitted.

6.5.4 Numerical solution of Lindhard-type TEs

An iterative scheme devised by Winterbon (1986), originally designed for the *gas* model TEs, is also applicable to *liquid* model TEs. In this scheme the term $F_l^n(E - T_n; E_s)$ that occurs in $L_l F_l^n(E; E_s)$ is expanded to first order using a Taylor series. Performing this expansion, the exact equation, $L_l F_l^n(E; E_s) = S_l^n(E; E_s)$, is approximated by an ordinary differential equation (ODE) given by

$$A_l(E)F_l^n(E; E_s) + B_l(E)\frac{\mathrm{d}F_l^n(E; E_s)}{\mathrm{d}E} = S_l^n(E; E_s), \qquad (6.52)$$

where

$$A_l(E) = N \int_0^{p_t} [1 - P_l(\cos \psi(E, p))] 2\pi p \, \mathrm{d}p,$$
$$B_l(E) = N \int_0^{p_t} T_n(E, p)P_l(\cos \psi(E, p)) 2\pi p \, \mathrm{d}p + NS_e(E). \qquad (6.53)$$

In the *liquid* model, the term $F_l^n(E - T_l - T_n; E_s)$ is expanded to first order using a Taylor series. The ODE that results has the same left-hand side as equation (6.52). In both models, the true solution $F_l^n(E; E_s)$ is represented by an approximate solution, $F_{l,0}^n(E; E_s)$, plus a correction term, $F_{l,1}^n(E; E_s)$. An initial guess for $F_{l,0}^n(E; E_s)$ is obtained by solving the first-order ODE given by

$$A_l(E)F_{l,0}^n(E; E_s) + B_l(E)\frac{\mathrm{d}F_{l,0}^n(E; E_s)}{\mathrm{d}E} = S_{l,0}^n(E; E_s), \qquad (6.54)$$

which is amenable to numerical solution by standard initial value techniques (Kahaner, Moler and Nash, 1989).

Initial value problems require that the integration start at a low energy E_0 (say 1 eV) at which the solution $F_l^n(E_0; E_s)$ is known, or can be approximated, and then integration is carried forward until a required maximum energy E_1 is reached. In this way, results for all energies between E_0 and E_1 are produced in a single program run. The initial value $F_l^n(E_0; E_s)$ can be estimated from analytic solutions to the LSS TE using power-law potentials (see Chapter 5).

Using the same numerical techniques, the correction term is obtained by solving another first-order ODE given by

$$A_l(E)F_{l,1}^n(E; E_s) + B_l(E)\frac{\mathrm{d}F_{l,1}^n(E; E_s)}{\mathrm{d}E} = S_{l,0}^n(E; E_s) - L_l F_{l,0}^n(E; E_s). \qquad (6.55)$$

The true solution is then obtained by repeatedly updating the approximate solution, i.e.

$$F_{l,0}^n(E; E_s) := F_{l,0}^n(E; E_s) + F_{l,1}^n(E; E_s), \qquad (6.56)$$

while obtaining new correction terms by using equation (6.55). In practice, up to ten iterations are necessary using the *liquid* model. The integrals for $A_l(E)$ and $B_l(E)$ need to be computed only once, at the beginning of a program, after the ion/target combination has been selected.

6.5.5 Coupling relations

Once the coefficients $F_l^n(E; E_s)$ have been computed, it is a simple matter to reconstruct the vertical and lateral moments of the distribution from equation (6.39). For implantation at normal incidence, vertical moments $\langle z^n \rangle$ can be obtained by setting $\eta = 1$ and lateral moments $\langle x^{2m} \rangle$ can be obtained by setting $\eta = 0$. Mixed moments of an arbitrary order can be obtained using the plane-source to point-source transformation (Berger and Spencer, 1959, Winterbon, Sigmund and Sanders, 1970)

$$\langle z^{n-2m} x^{2m} \rangle = ({}^n C_{2m})^{-1} \sum_{l=0}^{n/2} (2n - 4l + 1) F_{n-2l}^n b_{nml}, \tag{6.57}$$

where

$$b_{nml} = \frac{1}{2^{n-2l}} \sum_{k=\max (m,l)}^{n/2} {}^k C_m \frac{(-1)^{k-l}(2n - 2l - 2k)!}{(k - l)!(n - l - k)!(n - 2k)!}. \tag{6.58}$$

Some results of equation (6.58) for moments up to order four are

$$\langle z^2 \rangle = F_0^2 + 5F_2^2, \quad \langle x^2 \rangle = F_0^2 - \frac{5}{2}F_2^2,$$

$$\langle z^3 \rangle = 3F_1^3 + 7F_3^3, \quad \langle zx^2 \rangle = F_1^3 - \frac{7}{2}F_3^3,$$

$$\langle z^4 \rangle = F_0^4 + 5F_2^4 + 9F_4^4, \quad \langle x^4 \rangle = F_0^4 - \frac{5}{2}F_2^4 + \frac{27}{8}F_4^4, \tag{6.59}$$

$$\langle z^2 x^2 \rangle = \frac{1}{3}F_0^4 + \frac{5}{12}F_2^4 - \frac{9}{2}F_4^4,$$

where the energy-dependence of the coefficients $F_m^n(E; E_s)$ has been omitted.

6.5.6 Implementation

A computer program called KUBBIC (Kent University Backward Boltzmann Implantation Code) that numerically solves Lindhard-type TEs, using either the *gas* or *liquid* models, has been written. All integral and ODE computations are carried out on a common logarithmic energy grid called the 'computation' grid. The Biersack two-parameter 'magic' formula is used to compute the functions T_n and ψ that occur in the integrals $A_l(E)$, $B_l(E)$ and $L_l F_{l,0}^n(E; E_s)$. A second energy grid called the 'printout' grid, which is generated independently of the 'computation' grid, defines the energies at which results are output. The integrals are computed using an adaptive Gaussian quadrature algorithm, due to Patterson (1968), from the NAG library. The ODEs are computed using a variable order, variable step Adams method, again from the NAG library. The algorithms used to compute the integrals found in the ODE source term and for computation of the ODE itself require interpolation in order to evaluate quantities at arbitrary energies between grid points. This is carried out by using the piecewise cubic Hermite polynomial algorithm of Fritsch and co-workers (1980, 1984). The final results are checked for convergence in terms of the number of points in the computation grid and the number of correction iterations (over the energy range of interest).

6.6 Spatial moments of implantation profiles

In this section, spatial moments that describe the projection of the rest distribution in one dimension (1-D) and in two dimensions (2-D) are defined.

6.6.1 Projections

Because (i) the target is infinite and homogeneous and (ii) all azimuthal scattering angles are equally probable, the rest distribution of ions from a point mono-directional source is radially symmetric in the plane normal to $\hat{u}$. Hence, the shape of the rest distribution (in three dimensions) is determined by two independent variables (instead of three). If $\hat{u}$ is arbitrary, then suitable variables are (Sanders, 1968) $|r|$ and the cosine of the angle between r and $\hat{u}$. Furthermore, if $\hat{u}$ is confined to a plane, then the rest distribution is uniquely described by its projection in 2-D on this plane. This projection (the 2-D profile) is called the line source response (LSR). The term 'line source' is used because the same shaped profile results if the point mono-directional source is replaced by a line mono-directional source on the y axis. If the 2-D profile is known, then a transform (Ashworth, Bowyer and Oven, 1991) can be used to obtain the distribution in three dimensions. The projection in 1-D can be obtained by experiment (Hofker *et al.*, 1975). For this reason, the 1-D profile is important in its own right. However, a single 1-D profile does not contain the necessary information to determine the corresponding 2-D profile. The 1-D projection on the axis normal to the target surface is called the projected range distribution or, more rigorously, the vertical projected range distribution. In this book, the vertical distribution is the 1-D projection on the z axis. The projected range distribution is also known as the plane source response because the same shaped profile results if the point mono-directional source is replaced by a mono-directional source that occupies the x–y plane. The 1-D projection onto an arbitrary axis tangential to the target surface is called the lateral projected range distribution. In this book, the lateral distribution is the 1-D projection onto the x-axis.

Let $F(x, y, z)$ represent the rest distribution. The 2-D profile, $F_{2p}(z, x)$, is defined by

$$F_{2p}(z, x) = \int_{-\infty}^{\infty} F(x, y, z) \, dy. \tag{6.60}$$

The 1-D profile on the z axis, $F_p(z)$, is defined by

$$F_p(z) = \int_{-\infty}^{\infty} F_{2p}(z, x) \, dx \tag{6.61}$$

and the 1-D profile on the x axis, $F_L(x)$, is defined by

$$F_L(x) = \int_{-\infty}^{\infty} F_{2p}(z, x) \, dz. \tag{6.62}$$

6.6.2 Moments about the origin

The nth-order moment about the origin of $F_p(z)$ is defined by

$$\langle z^n \rangle = \int_{-\infty}^{\infty} z^n F_p(z) \, dz \tag{6.63}$$

and the mth-order moment about the origin of $F_L(x)$ is defined by

$$\langle x^m \rangle = \int_{-\infty}^{\infty} x^m F_L(x) \, dx. \tag{6.64}$$

The 2-D profile, $F_{2p}(z, x)$, is partly described by its marginal density functions $F_p(z)$ and $F_L(x)$. However, an accurate description of the 2-D profile requires the mixed moments defined by

$$\langle z^n x^m \rangle = \int_{-\infty}^{\infty} \int_{-\infty}^{\infty} z^n x^m F_{2p}(z, x) \, dz \, dx. \tag{6.65}$$

6.6.3 Vertical moments about $\langle z \rangle$

The first moment, $\langle z \rangle$, determines the position of the distribution. It is called the mean projected range. It is difficult to relate the moments $n \geq 2$, measured about the origin, to basic properties of the profile. The nth moment about $\langle z \rangle$, called the nth central moment, is defined by

$$\langle z_c^n \rangle = \int_{-\infty}^{\infty} (z - \langle z \rangle)^n F_p(z) \, dz \tag{6.66}$$

(where it is necessary to distinguish the moments $\langle z^n \rangle$ from the central moments $\langle z_c^n \rangle$; the moments about the origin are called 'non-central' moments.) The second central moment, called the variance, is a measure of the width of the profile. The square root of the variance is called the vertical standard deviation and is denoted by the symbol σ_z. The relation between central and non-central moments is (Spiegel, 1975)

$$\langle z_c^n \rangle = \sum_{j=0}^{n} {}^n C_j (-1)^{n-j} \langle z^j \rangle \langle z \rangle^{n-j}. \tag{6.67}$$

To specify the shape of the profile it is necessary to consider the third and higher moments. The moments that specify the shape of a profile must be invariant under linear transformation, i.e. independent of the position and width of the profile. Moments that meet this criterion are the normalised (or dimensionless) central moments defined by

$$\langle z_{cn}^n \rangle = \int_{-\infty}^{\infty} \left(\frac{z - \langle z \rangle}{\sigma_z} \right)^n F_p(z) \, dz. \tag{6.68}$$

The third normalised central moment is called the vertical skewness (Spiegel, 1975) and is denoted by the symbol γ_z. The skewness is a measure of the asymmetry of the profile. A negative skewness corresponds to a profile with a longer tail to the left of the peak and a positive skewness corresponds to a profile with a longer tail to the right of the peak. For a symmetric distribution the skewness is zero. The fourth normalised central moment is called the vertical kurtosis (Spiegel, 1975), and is denoted by the symbol β_z. The kurtosis is a measure of the degree of peakedness of the profile. The shape of the normal distribution is defined by the values $\gamma_z = 0$ and $\beta_z = 3$. A kurtosis greater than 3 corresponds to a peak that is sharper than the peak of the normal distribution whereas, conversely, a kurtosis less than 3 corresponds to a peak that is shallower. The smallest value of kurtosis that defines a physical implantation profile is 1.8 (Ashworth, Bowyer and Oven, 1991).

6.6.4 Lateral moments

If the implantation is performed along the z axis, then the rest distribution is symmetric about the z axis. In this case, the profile $F_L(x)$ is symmetric and, consequently, the moments $\langle x^m \rangle$, with m odd, are zero. A measure of the width of the lateral profile is the lateral standard deviation, $\sigma_x = \langle x^2 \rangle^{1/2}$ and the normalised central moments are defined by

$$\langle x_n^m \rangle = \int_{-\infty}^{\infty} \left(\frac{x}{\sigma_x} \right)^m F_L(x) \, dx, \qquad (6.69)$$

where m is even. The fourth normalised central moment is called the lateral kurtosis and is denoted by the symbol β_x.

6.6.5 Mixed moments

The normalised central mixed moments, $\langle z_{cn}^n x_n^m \rangle$, of the 2-D profile are defined by

$$\langle z_{cn}^n x_n^m \rangle = \int_{-\infty}^{\infty} \int_{-\infty}^{\infty} \left(\frac{z - \langle z \rangle}{\sigma_z} \right)^n \left(\frac{x}{\sigma_x} \right)^m F_{2p}(z, x) \, dz \, dx. \qquad (6.70)$$

The mixed moments with $n = 0$ reduce to the lateral moments and the mixed moments with $m = 0$ reduce to the vertical moments. In this case, the moments up to order four that describe the 2-D profile are

$$\langle z \rangle, \ \sigma_z, \ \gamma_z, \ \beta_z, \ \sigma_z, \ \beta_x, \ \langle z_{cn} x_n^2 \rangle \text{ and } \langle z_{cn}^2 x_n^2 \rangle. \qquad (6.71)$$

Henceforth, the moments in (6.71) will be called 'the first eight moments'.

6.7 Generating profiles in one and two dimensions

The density functions of one co-ordinate, constructed from moments, can be divided coarsely into two classes: *single density functions* and *assembled density functions*. It is found that a single density function can only model a restricted variety of observed distributions. To this end, families of density functions (also known as *systems of frequency curves*) have been devised to model a wide variety of observed distributions. Two well-established families are the Pearson and Johnson families (Johnson and Kotz, 1970). The best known examples of assembled density functions, the Edgeworth and Gram–Charlier series (Johnson and Kotz, 1970), cover a wide range of observed distributions. However, such series, though easy to construct, result in densities with negative excursions and oscillatory behaviour (which is non-physical).

Two broad classes of 2-D density functions can be identified: *bi-variate analogues of 1-D single density functions* and *density functions comprising the product of two 1-D density functions*. A 2-D analogue of the Pearson system has been derived by van Uven (Johnson and Kotz, 1972). Owing perhaps to their complexity, the application to ion-implantation of bi-variate analogues is hard to find. It has been remarked that the van Uven densities depend on a single mixed moment (Winterbon, 1986). The mixed moments are important because they contain information describing the correlation between vertical and lateral motion of the ion. By computing the first eight moments, the third- and fourth-order mixed moments are computed. It is desirable, therefore, to choose density functions that maximise this information. The second class, in contrast to the first, has a well-documented history in ion-implantation studies. In the first 2-D model, presented by Furukawa, Matsumara and Ishiwara (1972), the 2-D profile $F_{2p}(z, x)$ was represented by the product of two Gaussians

$$F_{2p}(z, x) = \frac{1}{\sigma_z (2\pi)^{\frac{1}{2}}} \exp\left[-\frac{1}{2}\left(\frac{z - \langle z \rangle}{\sigma_z}\right)^2\right] \frac{1}{\sigma_x (2\pi)^{\frac{1}{2}}} \exp\left[-\frac{1}{2}\left(\frac{x}{\sigma_x}\right)^2\right]. \quad (6.72)$$

This model has been enhanced (Ryssel and Hoffman, 1983, Ashworth and Moulavi-Kakhki, 1985) by replacing the Gaussian in the vertical direction by the 1-D profile, $F_p(z)$:

$$F_{2p}(z, x) = F_p(z)\frac{1}{\sigma_x (2\pi)^{\frac{1}{2}}} \exp\left[-\frac{1}{2}\left(\frac{x}{\sigma_x}\right)^2\right]. \quad (6.73)$$

Further sophistication has been added by Giles and Gibbons (1985), who, in addition to replacing the vertical Gaussian by $F_p(z)$, replaced the lateral Gaussian by a general lateral range density function, $F_L(x)$, to give

$$F_{2p}(z.x) = F_p(z)F_L(x). \quad (6.74)$$

Neither of these enhancements were proposed in the context of moments solutions and, in the latter case, $F_p(z)$ and $F_L(x)$ were obtained directly from numerical solution of the Forward Boltzmann Transport Equation. Nevertheless, in both cases the application of density functions determined by moments is clear. For example, $F_p(z)$ can be replaced by a curve determined from the vertical moments $\langle z \rangle$, σ_z, γ_z and β_z and $F_L(z)$ can be replaced by a symmetrical curve determined from the lateral moments σ_x and β_x. In spite of their degree of sophistication, such models (based on the product of two *independent* marginal density functions), though convenient, are flawed because they neglect correlation between the vertical and lateral coordinates. Ashworth and Oven (1986) incorporated this correlation by introducing a parameter called the depth-dependent lateral standard deviation, $\sigma_x(z)$, and used it in a generalised version of equation (6.73):

$$F_{2p}(z, x) = F_p(z) \frac{1}{\sigma_x(z)(2\pi)^{\frac{1}{2}}} \exp\left[-\frac{1}{2}\left(\frac{x}{\sigma_x(z)}\right)^2\right]. \tag{6.75}$$

Using this model, Oven and Ashworth (1987) demonstrated the inadequacy of the depth-independent models. In order to describe 2-D profiles obtained from MC simulations, Hobler, Langer and Selberherr (1987) presented a model incorporating the depth-dependent lateral standard deviation and the depth-dependent lateral kurtosis, $\beta_x(z)$. They used a generalisation of equation (6.74):

$$F_{2p}(z, x) = F_p(z)F_L(x, z), \tag{6.76}$$

where $F_L(x, z)$, the lateral density at a depth z, is determined from $\sigma_x(z)$ and $\beta_x(z)$. The lateral density at a depth z has been formally identified, in the notation of statistics, as the conditional density function, $f_c(x|z)$ (Ashworth, Bowyer and Oven, 1991). Suitable forms for $f_c(x|z)$ are the symmetric Pearson and Johnson types.[1]

It must be remembered that the TE solver described produces depth-independent lateral moments and that information to determine depth-dependent lateral moments is incorporated in the mixed moments. By using analytical models for the depth-dependent lateral moments it is possible to derive equations relating the model coefficients to the vertical, lateral and mixed moments.

A quantity directly related to the LSR is the distribution of ions under an ideal mask. Oven, Ashworth and Bowyer (1992) have shown that analytical models in the form given by equation (6.76) can be convoluted in order to obtain closed form expressions for this distribution.

6.7.1 The Pearson family of frequency curves

If $f(y)$ is a density function, where $y = z - \langle z \rangle$, the Pearson family of frequency curves has the design goal to force $df(y)/dy = 0$ at two points on the density function. This requirement is satisfied, in a general manner, by $df(y)/dy = f(y)(y - a)/F(y)$, where $F(y)$ is a general function of y. It can been seen, therefore, that, when $f(y)$ is zero (e.g. at the beginning of the curve) or when $y = a$, defining the position of the mode, then $df(y)/dy$ will be equal to zero. The Pearson family is defined, when $F(y)$ is expanded by

---

[1] For $\beta_x(z) < 3.0$, the modified Gaussian (Hobler, Langer and Selberherr, 1987) is an alternative form.

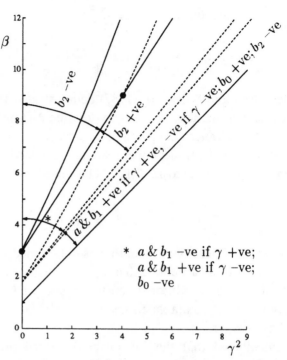

Fig. 6.4 Regions and lines on the β–γ^2 plane determined by the changes in sign of the Pearson coefficients $a = b_1$, b_0 and b_2, after Ashworth, Oven and Mundin (1990).

Maclaurin's theorem in powers of y, to give the differential equation

$$(b_0 + b_1 y + b_2 y^2 + \ldots)\frac{\mathrm{d}f(y)}{\mathrm{d}y} = (y - a)f(y). \tag{6.77}$$

Multiplying each side of equation (6.77) by y^n and integrating throughout (the left-hand side by parts, with $\mathrm{d}f(y)/\mathrm{d}y$ treated as one part, and the right-hand side treated as the sum of two functions) produces a recurrence relation (Elderton and Johnson, 1969). Setting $n = 0, 1, 2, 3$ gives a linear system containing the curve parameters a, b_0, b_1 and b_2, the zeroth-order moment and the first four non-central moments of $f(y)$. The linear system is conveniently solved by setting the zeroth-order moment equal to one (normalising the density to one) and setting the first moment equal to zero (placing the mean at the origin) to give three unique parameters in terms of three central moments. Using the dimensionless third and fourth moments, the curve parameters (called the Pearson coefficients) are given by

$$a = b_1 = -\gamma_z \sigma_z (\beta_z + 3)C,$$
$$b_0 = -\sigma_z^2(4\beta_z - 3\gamma_z^2)C, \tag{6.78}$$
$$b_2 = -(2\beta_z - 3\gamma_z^2 - 6)C,$$

where

$$C = [2(5\beta_z - 6\gamma_z^2 - 9)]^{-1}. \tag{6.79}$$

The equations (6.78) allow the mapping of regions and lines on the β–γ^2 plane which are determined by the changes in the sign of the Pearson coefficients as shown in Figure 6.4. The form of solution of Pearson's differential equation depends upon the nature of the roots of the equation

$$b_0 + b_1 y + b_2 y^2 = 0. \tag{6.80}$$

The full classification (Johnson and Kotz, 1970), see Figure 6.5, results in domains of validity and equations for three main types, four main transitional types and six other types. The shapes of various Pearson curves have three distinct classifications: bell-shaped, J-shaped (and twisted J-shaped) and U-shaped. Only bell-shaped curves are applicable to amorphous target profiles. It is readily shown from Pearson's differential equation that, for a maximum, $b_0 + b_1^2(1 + b_2) < 0$. This can be reformulated in terms of an inequality between β and γ^2 (Ashworth, Oven and Mundin, 1990):

$$\beta > \frac{9\{(6\gamma^2 + 5) + [\frac{9}{16}\gamma^6 + 8\gamma^4 + 25(\gamma^2 + 1)]^{\frac{1}{2}}\}}{50 - \gamma^2}, \tag{6.81}$$

with an additional constraint that $\gamma^2 < 50$. In this case, bell-shaped curves result from a region on the β–γ^2 plane bounded to the left by the vertical line $\gamma = 0$ and to the right by the line given by the inequality (6.81). This is the shaded area shown in Figure 6.5 which, from the figure, indicates that bell-shaped curves lie in a region covered by the main Pearson types I–VII (including the Gaussian). By implication, therefore, any one of the main Pearson types may be selected according to the third- and fourth-order moments obtained from implantation into an infinite monolayer. This simple conclusion contrasts with many conflicting reports about the applicability (or not) of individual Pearson types. Expressions from which the correct Pearson type may be selected are

Gaussian $\gamma = 0, \beta = 3$,
Type I $\quad \gamma \neq 0, (\gamma^2 + 1) \leq \beta < (3 + \frac{3}{2}\gamma^2), \Lambda \neq 0, C \neq \infty$,
Type II $\quad \gamma = 0, 1 \leq \beta < 3$,
Type III $\quad \gamma \neq 0, \beta = (3 + \frac{3}{2}\gamma^2)$,
Type IV $\quad 0 < \gamma^2 < 32, \beta > [39\gamma^2 + 48 + 6(\gamma^2 + 4)^{3/2}]/(32 - \gamma^2)$,
Type V $\quad 0 < \gamma^2 < 32, \beta = [39\gamma^2 + 48 + 6(\gamma^2 + 4)^{3/2}]/(32 - \gamma^2)$,
Type VI $\quad \gamma \neq 0, (3 + \frac{3}{2}\gamma^2) < \beta < [39\gamma^2 + 48 + 6(\gamma^2 + 4)^{3/2}]/(32 - \gamma^2)$,
$\quad\quad\quad\quad \Lambda \neq 0$,
Type VII $\quad \gamma = 0, \beta > 3$,

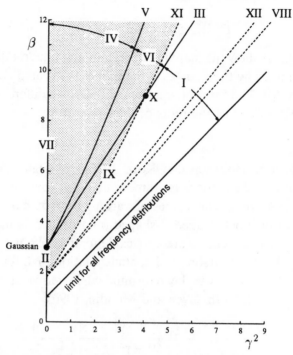

Fig. 6.5 Pearson types and domains of validity classified on the β–γ^2 plane, after Ashworth, Oven and Mundin (1990).

$$(6.82)$$

where

$$\Lambda = \gamma^2(\beta + 3)^2(8\beta - 9\gamma^2 - 12) - 4(4\beta - 3\gamma^2)(5\beta - 6\gamma^2 - 9)^2. \qquad (6.83)$$

Approximate relationships between β and γ^2 for the transition types can be derived by solving the equation $\Lambda = 0$ (Ashworth, Oven and Mundin, 1990). The main Pearson types have the following equations.

Gaussian

$$f(z) = K \exp\left(\frac{(z - \langle z \rangle)^2}{2b_0}\right) \quad (b_1 = b_2 = 0). \qquad (6.84)$$

Type I

$$f(z) = K(z - \langle z \rangle + A_1)^{M_1}(A_2 + \langle z \rangle - z)^{M_2}, \quad A_2 \geq (z - \langle z \rangle) \geq -A_1 \qquad (6.85)$$

in which

$$M_1 = \frac{A_1 + b_1}{b_2(A_1 + A_2)}, \quad M_2 = \frac{A_2 - b_1}{b_2(A_1 + A_2)}, \qquad (6.86)$$

where

$$A_{1,2} = \pm\frac{1}{2}\left\{\frac{b_1}{b_2} \pm \left[\left(\frac{b_1}{b_2}\right)^2 - \frac{4b_0}{b_2}\right]^{1/2}\right\}$$

or   (6.87)

$$A_{1,2} = \pm\frac{1}{2}\left\{\frac{b_1}{b_2} \mp \left[\left(\frac{b_1}{b_2}\right)^2 - \frac{4b_0}{b_2}\right]^{1/2}\right\}$$

such that A_1 and A_2 are both positive.

Type II

$$f(z) = K(A^2 - (z - \langle z \rangle)^2)^m, \quad A \ge (z - \langle z \rangle) \ge -A \; (b_1 = 0) \qquad (6.88)$$

in which $A = (-b_0/b_2)^{1/2}$ and $m = 1/(2b_2)$.

Type III

$$f(z) = K\left(1 + \frac{z - \langle z \rangle}{A}\right)^p \exp\left[-q(z - \langle z \rangle)\right]$$

with   (6.89)

$$1 + \frac{(z - \langle z \rangle)}{A} \ge 0 \; (b_2 = 0)$$

in which $A = (p+1)/q$, $q = -1/b_1$ and $p = -(1 + b_0/b_1^2)$.

Type IV

$$f(z) = K\left[1 + \left(\frac{z - \langle z \rangle}{A} - \frac{n}{r}\right)^2\right]^{-m} \exp\left[-n \arctan\left(\frac{z - \langle z \rangle}{A} - \frac{n}{r}\right)\right] \qquad (6.90)$$

in which $r = 2(m - 1)$, $n = -rb_1(4b_0b_2 - b_1^2)^{-1/2}$, $m = -1/(2b_2)$ and $A = mrb_1/n$.

Type V

$$f(z) = K\left(1 + \frac{z - \langle z \rangle}{A}\right)^{-p} \exp\left[\frac{A(2 - p)}{z - \langle z \rangle + A}\right]$$

with   (6.91)

$$1 + \frac{(z - \langle z \rangle)}{A} \ge 0 \; (b_1^2 = 4b_0b_2)$$

in which $p = -1/b_2$ and $A = b_1/(2b_2)$.

Type VI

$$f(z) = K \left(1 + \frac{z - \langle z \rangle}{A_1}\right)^{-q_1} \left(1 + \frac{z - \langle z \rangle}{A_2}\right)^{q_2}, \quad 1 + \frac{(z - \langle z \rangle)}{A_2} \geq 0 \qquad (6.92)$$

in which

$$q_1 = -\frac{A_1 + b_1}{b_2(A_1 - A_2)}, q_2 = -\frac{A_2 + b_1}{b_2(A_1 - A_2)}, \qquad (6.93)$$

where

$$A_{1,2} = \frac{1}{2} \left\{ \frac{b_1}{b_2} \pm \left[\left(\frac{b_1}{b_2}\right)^2 - \frac{4b_0}{b_2} \right]^{1/2} \right\}. \qquad (6.94)$$

Type VII

$$f(z) = K(A + (z - \langle z \rangle)^2)^{-m} \quad (b_1 = 0) \qquad (6.95)$$

in which $A = b_0/b_2$ and $m = -1/(2b_2)$. In these equations K is a normalisation constant, which can be determined analytically in all but one case: type IV. However, no difficulty is encountered in determining this constant by numerical integration.

The conditional density function, $f_c(x|z)$, can be represented by the symmetrical Pearson types II, VII and Gaussian which all have zero skewness. A generalisation of Pearson's differential equation that incorporates depth-dependent parameters is

$$\frac{\mathrm{d}f_c(x|z)}{\mathrm{d}x} = \frac{xf_c(x|z)}{b_0(z) + b_2(z)x^2} \qquad (6.96)$$

in which the functions $b_0(z)$ and $b_2(z)$ are related to the second and fourth depth-dependent lateral moments of the distribution through the relationships

$$\begin{aligned} b_0(z) &= -\sigma_x^2(z)[1 + 3b_2(z)], \\ b_2(z) &= [3 - \beta_x(z)]/(5\beta_x(z) - 9), \end{aligned} \qquad (6.97)$$

where the depth-dependent lateral variance is defined as

$$\sigma_x^2(z) = \frac{1}{\sigma_x^2} \int_{-\infty}^{\infty} x^2 f_c(x|z) \, \mathrm{d}x \qquad (6.98)$$

and the depth-dependent lateral kurtosis is defined as

$$\beta_x(z) = \frac{1}{\sigma_x^4(z)} \int_{-\infty}^{\infty} x^4 f_c(x|z) \, \mathrm{d}x. \qquad (6.99)$$

On solving equation (6.96), the (normalised) conditional density functions are (Ashworth, Bowyer and Oven, 1991) as follows.

Type II

$$f_c(x|z) = A(z)^{-(2p(z)+1)}(A^2(z) - x^2)^{p(z)}/B\left(\frac{1}{2}, p(z) + 1\right) \qquad (6.100)$$

in which $B(a,b)$ is the beta function, $A(z) = [-b_0(z)/b_2(z)]^{1/2}$ and $p(z) = 1/[2b_2(z)]$.

Type VII

$$f_c(x|z) = C(z)^{q(z)-\frac{1}{2}}(C(z) + x^2)^{-q(z)}/B\left(\frac{1}{2}, q(z) - \frac{1}{2}\right) \qquad (6.101)$$

in which $C(z) = b_0(z)/b_2(z)$ and $q(z) = -1/[2b_2(z)]$.

Gaussian

$$f_c(x|z) = (-2\pi b_0(z))^{-1/2} \exp(x^2/(2b_0(z))). \qquad (6.102)$$

The above Pearson types are either bell-shaped or U-shaped (in x) with $f_c(x|z) = $ constant as the boundary between the two. As before, bell-shaped distributions must be guaranteed. This means, by differentiation of equation (6.96), that $b_0(z) < 0$. Hence, by equations (6.97), the depth-dependent lateral kurtosis must be such that $\beta_x(z) > 1.8$. In this case Pearson type II is chosen for $1.8 < \beta_x(z) < 3$. The Gaussian is chosen when $\beta_x(z) = 3$ and Pearson type VII when $\beta_x(x) > 3$. An Abel-type integral transform equation has been used to obtain modified types II and VII suitable for the representation of the rest distribution in three dimensions. An analysis (which is not confined to Pearson frequency curves), contained in Ashworth, Bowyer and Oven (1991), proves that $\beta_x(z) \geq 2$ for bell-shaped conditional densities in 3-D. This suggests that the lower bound for $\beta_x(z)$, to be encountered for amorphous target problems, is 2.0 and not 1.8.

6.8 Models for depth-dependent lateral moments

In this section analytical models are sought for the depth-dependent lateral moments $\sigma_x(z)$ and $\beta_x(z)$. The models must be such that the coefficients can be obtained from depth-independent spatial moments and, in particular, the mixed moments. The following double-integral defines the depth-independent moments of the 2-D profile $F_{2p}(z, x)$:

$$\langle z_{cn}^j x_n^k \rangle = \int_{-\infty}^{\infty} \int_{-\infty}^{\infty} z_{cn}^j x_n^k F_{2p}(z, x) \, dx \, dz, \qquad (6.103)$$

where $z_{\mathrm{cn}} = (z - \langle z \rangle)/\sigma_z$ and $x_{\mathrm{n}} = x/\sigma_x$. Introducing $F_{2\mathrm{p}}(z, x) = F_{\mathrm{p}}(z)f_{\mathrm{c}}(x|z)$ into equation (6.103) gives

$$\langle z_{\mathrm{cn}}^j x_{\mathrm{n}}^k \rangle = \int_{-\infty}^{\infty} z_{\mathrm{cn}}^j F_{\mathrm{p}}(z) \left(\int_{-\infty}^{\infty} x_{\mathrm{n}}^k f_{\mathrm{c}}(x|z)\,\mathrm{d}x \right) \mathrm{d}z. \qquad (6.104)$$

The inner integral in equation (6.104) is the definition of the kth-order normalised depth-dependent lateral moment $\langle x_{\mathrm{n}}^k(z) \rangle$. At this stage the $\langle x_{\mathrm{n}}^k(z) \rangle$ should be replaced by a suitable model (chosen for all k) or models (for specific k). Setting $j = 0, 1, 2...$ and $k = 2, 4, 6...$ provides a set of integral equations from which the model coefficients can be obtained. The second and fourth depth-dependent lateral moments are required in order to construct $f_{\mathrm{c}}(x|z)$. Setting $k = 2$ and $k = 4$, in equation (6.104), gives integral equation sets for the moments $\langle x_{\mathrm{n}}^2(z) \rangle$ and $\langle x_{\mathrm{n}}^4(z) \rangle$:

$$\langle z_{\mathrm{cn}}^j x_{\mathrm{n}}^2 \rangle = \int_{-\infty}^{\infty} z_{\mathrm{cn}}^j F_{\mathrm{p}}(z) \langle x_{\mathrm{n}}^2(z) \rangle\,\mathrm{d}z, \quad j = 0, 1, 2..., \qquad (6.105)$$

$$\langle z_{\mathrm{cn}}^j x_{\mathrm{n}}^4 \rangle = \int_{-\infty}^{\infty} z_{\mathrm{cn}}^j F_{\mathrm{p}}(z) \langle x_{\mathrm{n}}^4(z) \rangle\,\mathrm{d}z, \quad j = 0, 1, 2... \,. \qquad (6.106)$$

The moments $\langle x_{\mathrm{n}}^2(z) \rangle$ and $\langle x_{\mathrm{n}}^4(z) \rangle$ are related to the moments $\sigma_x(z)$ and $\beta_x(z)$ by using

$$\langle x_{\mathrm{n}}^2(z) \rangle = \frac{\sigma_x^2(z)}{\sigma_x^2}\,, \quad \langle x_{\mathrm{n}}^4(z) \rangle = \beta_x(z)[\langle x_{\mathrm{n}}^2(z) \rangle]^2. \qquad (6.107)$$

A simple quadratic model for depth-dependent lateral variance was proposed by Lorenz, Krüger and Barthel (1989) in which the integral equations were reduced to a system of linear equations. They used a perturbation technique which resulted in approximate equations for the model coefficients. However, Ashworth, Oven and Bowyer (1991) have shown that this mathematical step is unnecessary when using a polynomial model. They show that *direct* substitution of the quadratic model into equation (6.105) results in a system of linear equations. This can be seen by substituting

$$\langle x_{\mathrm{n}}^k(z) \rangle = \int_{-\infty}^{\infty} x_{\mathrm{n}}^k f_{\mathrm{c}}(x|z)\,\mathrm{d}x = B_k + C_k z_{\mathrm{cn}} + D_k z_{\mathrm{cn}}^2$$
$$\equiv Q_k(z) \qquad (6.108)$$

in equation (6.104) to obtain

$$\langle z_{\mathrm{cn}}^j x_{\mathrm{n}}^k \rangle = B_k \int_{-\infty}^{\infty} z_{\mathrm{cn}}^j F_{\mathrm{p}}(z)\,\mathrm{d}z + C_k \int_{-\infty}^{\infty} z_{\mathrm{cn}}^{j+1} F_{\mathrm{p}}(z)\,\mathrm{d}z$$
$$+ D_k \int_{-\infty}^{\infty} z_{\mathrm{cn}}^{j+2} F_{\mathrm{p}}(z)\,\mathrm{d}z, \qquad (6.109)$$

where B_k, C_k and D_k are the model coefficients. The integrals in equation (6.109) are the central moments defined by the standardised variable z_{cn}. For $j = 0 \dots 4$ they are given by

$$1 = \int_{-\infty}^{\infty} F_p(z) \, dz, \quad 0 = \int_{-\infty}^{\infty} z_{cn} F_p(z) \, dz, \quad 1 = \int_{-\infty}^{\infty} z_{cn}^2 F_p(z) \, dz,$$
$$\gamma_z = \int_{-\infty}^{\infty} z_{cn}^3 F_p(z) \, dz, \quad \beta_z = \int_{-\infty}^{\infty} z_{cn}^4 F_p(z) \, dz. \tag{6.110}$$

Substituting equations (6.110) into the set of three equations formed when $j = 0, 1, 2$ in equations (6.109) results in the following simultaneous equations:

$$\langle x_n^k \rangle = B_k + D_k,$$
$$\langle z_{cn} x_n^k \rangle = C_k + D_k \gamma_z, \tag{6.111}$$
$$\langle z_{cn}^2 x_n^k \rangle = B_k + C_k \gamma_z + D_k \beta_z.$$

Solving for the model coefficients gives

$$B_k = \langle x_n^k \rangle - D_k,$$
$$C_k = \frac{\langle z_{cn} x_n^k \rangle (1 - \beta_z) + \gamma_z (\langle z_{cn}^2 x_n^k \rangle - \langle x_n^k \rangle)}{1 + \gamma_z^2 - \beta_z},$$
$$D_k = \frac{\langle x_n^k \rangle - \langle z_{cn}^2 x_n^k \rangle + \gamma_z \langle z_{cn} x_n^k \rangle}{1 + \gamma_z^2 - \beta_z}. \tag{6.112}$$

Setting $k = 2$ ($\langle x_n^2 \rangle = 1$) gives the coefficients for the model $\sigma_x(z) = \sigma_x [Q_2(z)]^{1/2}$ and setting $k = 4$ ($\langle x_n^4 \rangle = \beta_x$) gives the coefficients for the model $\beta_x(z) = Q_4(z)/[Q_2(z)]^2$. Ashworth, Oven and Bowyer (1991) have compared the $\sigma_x(z)$ constructed from the exact model coefficients (6.112) and those constructed in Lorenz, Krüger and Barthel (1989) using the approximate coefficients, with $\sigma_x(z)$ from high-resolution MC data. Their results clearly demonstrate the superiority of the exact model coefficients.

The model for $\beta_x(z)$ requires fifth- and sixth-order moments (in addition to the first eight moments), namely, the mixed moments $\langle z_{cn} x_n^4 \rangle$ and $\langle z_{cn}^2 x_n^4 \rangle$. Without additional moments simplistic models for $\beta_x(z)$ are necessary. One such model, $\beta_x(z) = \beta_x$, has been employed by Lorenz, Krüger and Barthel (1989). This excludes any correlation between the vertical and lateral motion of the ion. An alternative model, $\beta_x(z) = \beta_{alt}$, which includes correlation, is derived by substitution of (Bowyer, 1993)

$$\langle x_n^4(z) \rangle = [Q_2(z)]^2 \beta_{alt} \tag{6.113}$$

into equation (6.106) with $j = 0$. This has the solution

$$\beta_{\text{alt}} = \frac{\beta_x}{B_2^2 + 2B_2D_2 + C_2^2 + 2C_2D_2\gamma_z + D_2^2\beta_z}. \tag{6.114}$$

It has been shown that the use of quadratic models for the moments $\langle x_n^2(z) \rangle$ and $\langle x_n^4(z) \rangle$ results in linear systems with simple solutions. However, this is not the general case. Logarithmic or exponential models provide no such simplification and must rely on optimisation techniques to obtain the model coefficients. In addition, the moments $\langle x_n^2(z) \rangle$ and $\langle x_n^4(z) \rangle$ are not directly the quantities of interest, these are $\sigma_x(z)$ and $\beta_x(z)$.

6.9 Comparison of TE and MC computer codes

6.9.1 Moments

Figures 6.6 (B ions implanted into a-Si) and 6.7 (As ions implanted into a-Si) compare the first eight moments generated by KUBBIC-92 and a parallel processor MC code based on TRIM (Ziegler, Biersack and Littmark, 1985). Both codes use the Biersack two-parameter 'magic' formula with the ZBL interatomic potential, the ZBL electronic loss procedure (Ziegler *et al.*, 1985) with the SCOEF88 atomic data table, $p_t = \pi^{-1/2}N^{-1/3}$ and $E_s = 5$ eV. Moments obtained from the MC code are the result of simulating 1 000 000 ion trajectories for energies up to 100 keV. The moments obtained at 300 keV are the result of 250 000 ions and at 1 MeV the result of 100 000 ions. The moments obtained using 250 000 ion trajectories guarantee the stability of the moments up to fourth order to better than 1% (Bowyer, 1993).

From Figures 6.6 and 6.7, the moments obtained from the TE and MC techniques show good agreement. For B ions ($Z_1 = 5$), the moments obtained from the *gas* and *liquid* models are in good agreement over the entire energy range. This is hardly surprising because the dominant stopping mechanism, in much of the slowing down, is electronic energy loss, which is common to both models. For As ions ($Z_1 = 33$), the moments obtained from the *gas* and *liquid* models diverge significantly at energies below about 10 keV. It has been shown (Bowyer, Ashworth and Oven, 1992) that, for low-energy heavy ions, this divergence is larger than that which occurs due to differences between the one- and two-parameter nuclear scattering mechanisms.

6.9.2 1-D profiles

In the quest for smaller device geometries, accurate modelling of the near surface region and of the profile tail becomes important. Accurate device

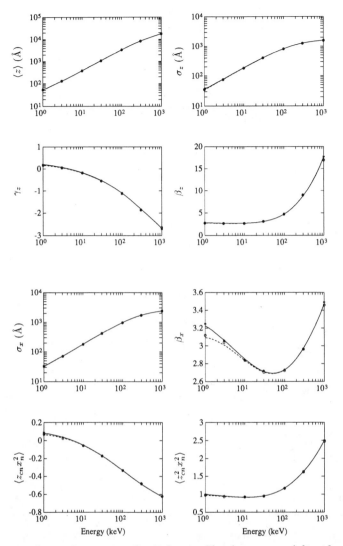

Fig. 6.6 Comparison of moments for B into a-Si using *gas* and *liquid* models with $E_s = 5$ eV and $p_t = \pi^{-1/2}N^{-1/3}$: o o o, MC with *gas* model; * * *, MC with *liquid* model; ———, KUBBIC-92, *gas* model; - - -, KUBBIC-92, *liquid* model.

modelling requires knowledge of the profile concentration over three or more orders of magnitude. Comparisons over such a profile dynamic range (PDR) require a logarithmic concentration scale. However, details of the profile normalisation and the general appreciation of shape are better suited to a linear concentration scale. For these reasons, profile comparisons should take place on both linear and logarithmic scales.

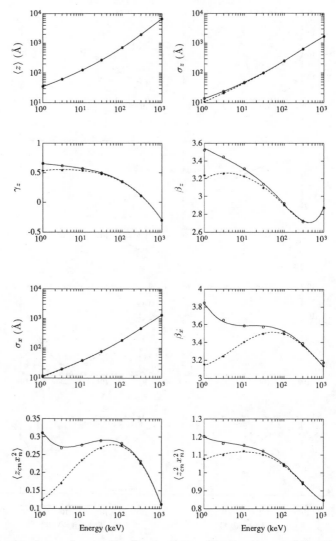

Fig. 6.7 Comparison of moments for As into a-Si using *gas* and *liquid* models with $E_s = 5$ eV and $p_t = \pi^{-1/2}N^{-1/3}$: o o o, MC with *gas* model; * * *, MC with *liquid* model; ——, KUBBIC-92, *gas* model; - - -, KUBBIC-92, *liquid* model.

Pearson and Johnson[1] curves, constructed using the (infinite) moments obtained from KUBBIC-92, are compared below with profiles obtained directly from MC simulations. To assess the change in profile shape due to ion reflection, MC simulations were performed using both real (semi-infinite) and infinite targets. To facilitate comparisons the areas under both the real

---

[1] The details of the construction of Johnson curves are beyond the scope of this chapter. The interested reader should consult Bowyer (1993).

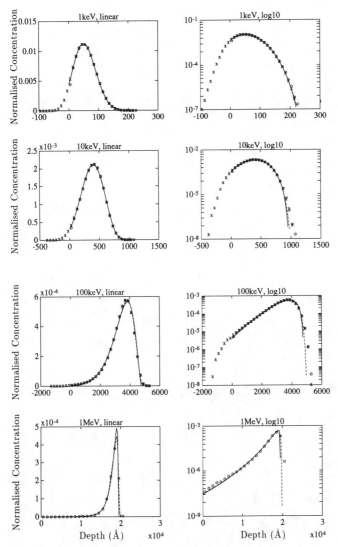

Fig. 6.8 Pearson and Johnson curves constructed from KUBBIC-92 moments compared with Monte Carlo profiles for B ions implanted into a-Si: o o o, MC profile using real target; x x x, MC profile using infinite target; ———, Pearson frequency curve; - - -, Johnson frequency curve.

and the infinite profiles were normalised over the right-hand half-plane (the real portion of the target).

Figures 6.8 and 6.9 show the comparisons performed for the ions B and As, respectively, implanted into a-Si at four energies: 1, 10 and 100 keV and 1 MeV. In both comparisons the *gas* transport model is employed with parameters $p_t = \pi^{-1/2}N^{-1/3}$ and $E_s = 5$ eV. The fits obtained using Pearson

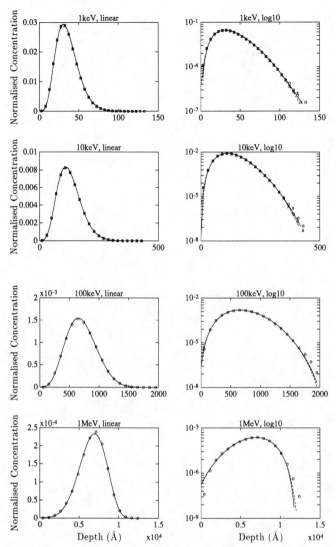

Fig. 6.9 Pearson and Johnson curves constructed from KUBBIC-92 moments compared with Monte Carlo profiles for As ions implanted into a-Si: o o o, MC profile using real target; x x x , MC profile using infinite target; ———, Pearson frequency curve; - - -, Johnson frequency curve.

and Johnson curves are, in the majority of cases, indistinguishable on a linear scale. The quality of fits on the linear scale, and over two or three orders of magnitude on the logarithmic scale, are good. The only discrepancies of note occur in each profile tail. In cases of significant surface concentration (B implants at 1 and 10 keV), comparisons of MC simulations using real

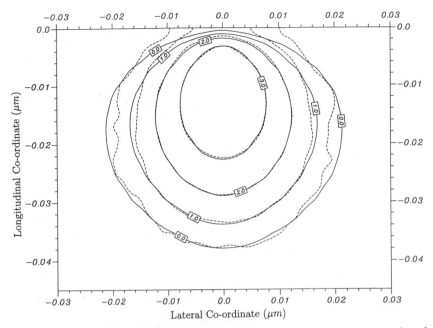

Fig. 6.10 A comparison of LSR for 10 keV As into a-Si: ———, construction from depth-independent moments using Pearson types I and VII, $Q_2(z)$ and β_x; - - -, rest distribution from MC simulation.

and infinite targets show that ion reflection has a negligible effect on the profile shape.

6.9.3 2-D profiles

The comparisons take the form of contour plots of $\log_{10}$ concentration. In each plot the concentration is normalised such that

$$\int_{-\infty}^{\infty} \int_{0}^{\infty} F_{2p}(z,x)\,dz\,dx = \int_{-\infty}^{\infty} \int_{0}^{\infty} F_p(z)f_c(x|z)\,dz\,dx = 1. \qquad (6.115)$$

Consecutive contour pairs represent a change in concentration of one order of magnitude. Implantation takes place at the coordinates $(x,z) = (0,0)$ in a direction downwards along the z axis. The z axis is termed the longitudinal co-ordinate and the x axis the lateral co-ordinate. The longitudinal co-ordinates are shown negative to be consistent with implantation downwards. The comparisons for 10 keV As implanted into a-Si are computed using the model $\beta_x(z) = \beta_x$ (Figure 6.10) and using the model $\beta_x(z) = \beta_{\text{alt}}$ (Figure 6.11). In both comparisons the *gas* transport model is employed with parameters $p_t = \pi^{-1/2}N^{-1/3}$ and $E_s = 5$ eV. The fits show that the model $\beta_x(z) = \beta_{\text{alt}}$ is closer to the MC simulation than is the model $\beta_x(z) = \beta_x$.

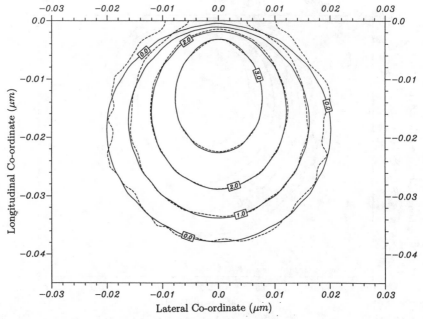

Fig. 6.11 Comparison of LSR for 10 keV As into a-Si: ———, construction from depth-independent moments using Pearson types I and VII, $Q_2(z)$ and β_{alt}; - - -, the rest distribution from MC simulation.

6.10 Additional remarks

Thus far the treatment of multi-element/multi-layer targets has not been discussed. In addition, it has been assumed that nuclear collisions are purely elastic.

A simple treatment of multi-element targets in both the TE and MC techniques (Sanders, 1968, Ziegler, Biersack and Littmark, 1985) uses the W. H. Bragg rule which states that the stopping power of a compound can be estimated by the linear combination of the stopping powers of the individual elements. The statistical interpretation of the Bragg rule is simply that the probability of an encounter with a target atom of given type is equal to its stoichiometric fraction, e.g. if the target is SiO_2, then oxygen atoms are encountered twice as frequently as silicon atoms. For a detailed discussion of the stopping in compounds see Ziegler and Manoyan (1988).

The treatment of multi-layer targets using the MC technique is straight-forward. In order to select the correct values of Z_2, m_2 and N, it is simply necessary to know, collision by collision, which layer the ion is in. Direct treatment of multi-layer structures using Lindhard-type TEs is impossible. However, knowledge of the rest distribution of ions as a function of implant energy, in each component of the target, provides the starting point for several multi-layer models (Ryssel and Biersack, 1986, Webb and Maydell, 1988).

Finally, the statistical transport model can be extended in order to incorporate inelastic nuclear scattering (in the manner described in Chapter 2). In particular, $\bar{\theta}$ is replaced by ϕ_m, given by equation (2.20), and T_n is replaced by T, given by equation (2.43).

References

Ashworth, D. G., Bowyer, M. D. J., and Oven, R. (1991). *J. Phys.* D **24** 1120.

Ashworth, D.G. and Moulavi-Kakhki, M. (1985). *Physica* B **129** 176.

Ashworth, D. G. and Oven, R. (1986). *J. Phys.* C **19** 5769.

Ashworth, D. G., Oven, R. and Bowyer, M. D. J. (1991). *Electron Lett.* **27** 1402.

Ashworth, D. G., Oven, R. and Mundin, B. (1990). *J. Phys.* D **23** 870.

Berger, M. J. and Spencer, L. V. (1959). *Phys. Rev.* **113** 408.

Biersack, J. P. (1982). *Z. Phys.* A **305** 95.

Bowyer, M. D. J. (1993). *Simulation of Ion Implantation in Amorphous Targets Using Moments Solutions of Transport Equations with Emphasis on Silicon Technology* PhD thesis, Electronic Engineering Laboratories, The University of Kent at Canterbury, UK.

Bowyer, M. D. J., Ashworth, D. G. and Oven, R. (1992). *J. Phys.* D **25** 1619.

Bowyer, M. D. J., Ashworth, D. G. and Oven, R. (1994). *J. Phys.* D **27** 2592.

Brice, D. K. (1971). *Rad. Eff. Defects Solids* **11** 227.

Duderstadt, J. J. and Martin, W. R. (1979). *Transport Theory*, Wiley, New York.

Eckstein, W. (1991). *Computer Simulation of Ion–Solid Interactions*, Springer-Verlag, Berlin.

Elderton, W. P. and Johnson, N. L. (1969) *Systems of Frequency Curves*, Cambridge University Press, London.

Fedder, S. and Littmark, U. (1981). *J. Appl. Phys.* **52** 4259.

Fritsch, F. N. and Butland, J. (1984). *SIAM J. Numer. Anal.* **5** 300.

Fritsch, F. N. and Carlson, R. E. (1980). *SIAM J. Numer. Anal.* **17** 238.

Furukawa, S., Matsumura, H. and Ishiwara, H. (1972). *Jap. J. Appl. Phys.* **11** 134.

Giles, M. D. and Gibbons, J. F. (1985). *J. Electrochem. Soc.* **132** 2476.

Hobler, G., Langer, E. and Selberherr, S. (1987). *Solid State Electronics* **30** 445.

Hofker, W. K., Oosthoek, D. P., Koeman, N. J. and De Grefte, H. A. M. (1975). *Rad. Eff. Defects Solids* **24** 223 (The PhD thesis of W. K. Hofker is contained in Philips Research Reports Supplements, 5-9, 1975.)

Johnson, N. L. and Kotz, S. (1970). *Distributions in Statistics: Continuous Univariate Distributions – 1*, Houghton Mifflin, Boston.

Johnson, N. L. and Kotz, S. (1972). *Distributions in Statistics: Continuous Multivariate Distributions*, Wiley, New York.

Kahaner, D., Moler, C. and Nash, S. (1989). *Numerical Methods and Software*, Prentice Hall Inc., New Jersey.

Knuth, D. E. (1981). *The Art of Computer Programming* (2nd edition), Addison Wesley, Reading, Massachusetts.

Lindhard, J., Scharff, M. and Schiøtt, H.E. (1963). *Kgl. Dansk. Vid. Selsk. Mat. Fys. Medd.* **33**.

Littmark, U. and Gras-Martí, A. (1978). *Appl. Phys.* **16** 247.

Lorenz, J., Krüger, W. and Barthel, A. (1989). Simulation of the lateral spread of implanted ions: theory *Proceedings of the Sixth International NASECODE Conference* Ed J. J. H. Miller, Boole Press, Dublin, 513–20.

Marsaglia, G. (1985). A current view of random number generators *Proceedings of the 16th Symposium, The Interface* Ed L. Billard, Elsevier, Amsterdam, 3–10.

Miyagawa, Y. and Miyagawa, S. (1983). *J. Appl. Phys.* **54** 7124.

Morgan, B. J. T. (1984). *Elements of Simulation*, Chapman and Hall, London.

Oven, R. and Ashworth, D. G. (1987). *J. Phys.* D **20** 642.

Oven, R., Ashworth, D. G. and Bowyer, M. D. J. (1992). *J. Phys.* D **25** 1235.

Oven, R., Bowyer, M. D. J. and Ashworth, D. G. (1993). *J. Phys.* C **5** 2157.

Patterson, T. N. L. (1968). *Math. Comp.* **22** 847.

Ryssel, H. and Biersack, J. P. (1986). *Ion Implantation Models for Process Simulation*, Ed W. J. Engl, Elsevier, North Holland, Amsterdam, ch 2, 31–69.

Ryssel, H. and Hoffman, K. (1983). *Ion Implantation* Ed. P. Antognetti, D. A. Antoniadis, R. W. Dutton and W. G. Oldham, Martinus Nijhoff, Den Haag, 125–79.

Sanders, J. B. (1968). *Can. J. Phys.* **46** 455.

Spiegel, M.R. (1975). *Schaum's Outline of Theory and Problems of Probability and Statistics*, McGraw-Hill, New York.

Spiegel, M. R. (1980). *Schaum's Outline Series: Mathematical Handbook of Formulas and Tables*, McGraw-Hill, New York.

Sze, S. M. (1988). *VLSI Technology* (2nd edition), McGraw-Hill, New York.

Webb, R. P. and Maydell, E. (1988). *Nucl. Instrum. Meth.* B **33** 117.

Winterbon, K. B. (1986). *Nucl. Instrum. Meth.* B **17** 193.

Winterbon, K. B., Sigmund, P. and Sanders, J. B. (1970). *Kgl. Dansk. Vid. Selsk. Mat. Fys. Medd.* **87**.

Ziegler J. F., Biersack, J. P. and Littmark U. (1985). The stopping and range of ions in solids *The Stopping and Ranges of Ions in Matter* Vol. 1 Ed J. F. Ziegler, Pergamon, New York.

Ziegler, J. F. and Manoyan, J. M. (1988). *Nucl. Instrum. Meth.* B **35** 215.

7

Binary collision algorithms

7.1 Introduction

The energetic interaction of a particle beam with a solid cannot be described fully by the path of a single projectile. The path a particle takes and the paths of the subsequent recoils are dependent upon the initial impact point on the surface. Thus, to get a clear description of the effects of particle interaction with a solid, many such paths must be followed. A typical ion beam experiment would entail the interaction of 10^{11}–10^{20} particles per cm^2 of the target.

Trajectory simulations obtain an ensemble – or set – of independent particle solid impact histories. Each history is followed from a different starting point on the solid to simulate the arrival of many particles at random points on the surface.

Conceptually the molecular dynamics (MD) simulation method (see Chapter 8) is the simplest and most complete simulation method to model the behaviour of a solid undergoing energetic particle bombardment; in particular, for calculating the displacement of particles in the solid during a single particle impact. In principle, the development of the ensuing collision cascade is followed chronologically in time as the energy of the ions propagates through the target system. The complexity comes from the solution of the many-body equations of motion which must be performed at successive time steps. The binary collision (BC) types of simulations – also often referred to as the event store method – are conceptually more complicated, although usually easier to program and, because of the simplifications of the models, execute much faster. There is, of course, the reliance that the assumptions made for the various simplifications to apply are upheld. It has often been found that, even when these are not upheld, the very nature of experimental observation of many particles has meant that the exact details of each single event are unimportant and general trends are often quite accurately described using these approximations.

The BC simulations are based, as their name suggests, on two-body collisions (see Chapter 2). This means that, instead of having to calculate the complete flight path of each particle during the collision cascade, their trajectories can be replaced by a set of straight-line paths. Each change of direction of the path is due to a binary collision; the angle through which the path changes and the energy exchanged during the collision are dependent upon the target material and ordering. In their simplest form BC simulations relax to Monte Carlo simulations in which only the flight path of the impinging ion is considered and the fate of the recoiling target atoms is ignored. A simple Monte Carlo algorithm for modelling low-dose ion implantation in amorphous materials is described in Chapter 6.

7.2 Collisions

The concept of a 'collision' exists in the context of the BC simulations primarily because, in the approximation used, the velocities of the colliding particles change discretely at these collision points (Robinson and Torrens, 1974). In the case of MD simulations there is no built-in concept of a collision. Indeed, because of the infinite nature of the interatomic potentials used, all particles are interacting with every other particle at all times. This would imply that particles are continuously 'colliding' with all others. However, not all of the interactions are 'significant' enough to affect the particles' motion. Often in MD simulations the interaction potential function is truncated to avoid the unnecessary calculation of insignificant interactions.

Figure 7.1 shows an MD study (Webb *et al.*, 1993) of the collision in a cascade initiated by a single 25 keV argon ion impact on a rhodium substrate. For the purposes of this study, a collision was defined as occurring if the combined magnitude of the forces on a particular target atom from the surrounding particles was greater than some threshold value, in this case 3×10^{-13} N. Thus low-energy, low-force interactions would not be registered as collisions. The collisions were then analysed to find out how many of the surrounding atoms were involved in the collision – in other words, how many particles contributed significantly to the collision. Figure 7.1 shows that the total number of collisions increases with time until the energy of the initial impact is spread so 'thinly' that any further interactions have insufficient energy to be greater than the threshold and hence are not registered as collisions. At this time the total number of collisions starts to decrease. For the BC model to be applied correctly we would like to be able to assume that

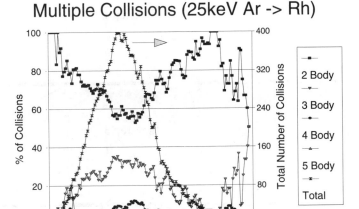

Fig. 7.1 Many-body collisions as a function of time for a 25 keV Ar impact on a rhodium crystal.

most of the collisions will be two-body ones. As we can see from Figure 7.1, in this particular case the percentage of two-body collisions decreases to only 60% at the peak of the activity of the cascade. A recent study (Chakarov *et al.*, 1995) indicated that, although the actual details of the cascade may not be simulated completely correctly, the propagation of energy is simulated correctly in these binary models.

7.3 Event store models

The BC algorithm results in a procedure which calculates a series of straight-line paths for each particle set in motion. These paths all have different starting times; compare Figures 7.2 and 7.3. Therefore, it can be quite complex to determine the interaction, if any, of interacting paths. In dilute, linear cascades the individual paths are assumed to be separated and all collisions to occur with the struck particle initially at rest. This results in a great saving in calculation time. The framework for this type of calculation is to follow a single path of collisions – or 'events' – storing the information of the collision partner so that this subsequent new set of paths can be followed at a later time in the calculation. The calculation builds up a tree structure of collision paths and follows each one to completion separately.

Robinson recently produced a version of the MARLOWE code (Robinson, 1989) that is capable of determining and correcting for the interaction of these straight-line paths. This version of MARLOWE code also includes the

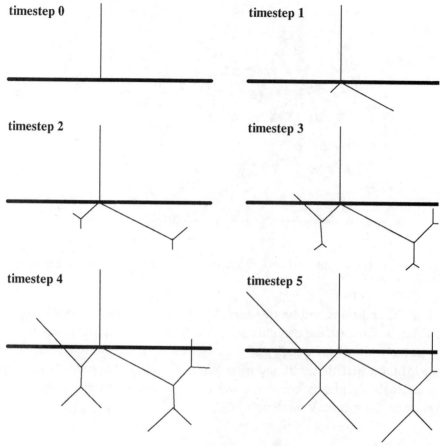

Fig. 7.2 The calculation-time behaviour of particle paths using a time-step model.

slowing down time of the particles. A time version (Vichev and Karpuzov, 1995) of the popular TRIM program (Biersack and Eckstein, 1984) has also been produced to allow for a better treatment of overlapping recoil paths in the investigation of dense cascades.

7.4 The genealogy of a binary collision program

7.4.1 Next event algorithms

A BC cascade simulation program constructs a series of straight-line paths from collision to collision, or event to event. A path is followed and the collision partners at each event are remembered until the energy of the particle in the original path is so low that it cannot cause further displacement. The procedure now is to recall and follow one of the sub-

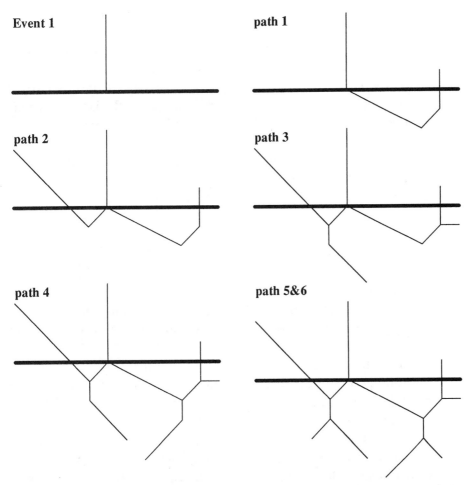

Fig. 7.3 The calculation-time behaviour of particle paths using event store logic.

events to its ultimate completion and so on. This is a simple matter of following a tree structure through all of its branches. There are a number of different algorithms by which the order of calculation of a specific set of paths may be calculated. In a system that totally ignores the interaction of paths the particular algorithm used should make no difference to the average result. However, differences can be found in the computational performance of the algorithm. In a system for which an attempt has been made to account for the interaction of paths, the particular algorithm used can be quite important to improve performance. Possible algorithms are as follows. A particle is followed until it has a collision. The energy of the original particle is shared between it and the particle it struck according to the two-body algorithm and then one of the following procedures is adopted.

1. Store both partners in a list, check through the list for the fastest particle and follow the fastest particle in the system.
2. Store both partners in a list, check through the list for the slowest particle and follow the slowest particle in the system.
3. Store the slowest particle in a list and follow the fastest partner from the collision; if this collision results in both partners having insufficient energy to make further displacements then the list is searched for the fastest particle.
4. Store the fastest partner in a list and follow the slowest partner from the collision; if the collision results in both partners having insufficient energy to make further displacements then the list is searched for the slowest particle.
5. It is possible to construct algorithms whereby the slowest of the collision pair is followed and the fastest particle from the list is chosen when the colliding pair have stopped, or the fastest of the collision pair and the slowest particle from the list. Both of these would form an inconsistent algorithm by mixing a choice of following the fastest or slowest particle and they have no obvious advantages over 3 and 4.
6. Always follow the same particle, storing the other collision particle in a list until the original particle has insufficient energy to cause further displacement when the *fastest* particle in the event list is followed exclusively.
7. Always follow the same particle, storing the other collision particle in a list until the original particle has insufficient energy to cause further displacement when the *slowest* particle in the event list is followed exclusively.
8. Always follow the same particle, storing the other collision particle in a list until the original particle has insufficient energy to cause further displacement when the *first* particle in the event list is followed exclusively.
9. Always follow the same particle, storing the other collision particle in a list until the original particle has insufficient energy to cause further displacement when the *last* particle in the event list is followed exclusively.
10. Always follow the other particle in the collision, storing the original particle in a list until the particle chosen has insufficient energy to cause further displacement when the *fastest* particle in the event list is followed.

11. Always follow the other particle in the collision, storing the original particle in a list until the particle chosen has insufficient energy to cause further displacement when the *slowest* particle in the event list is followed.

12. Always follow the other particle in the collision, storing the original particle in a list until the particle chosen has insufficient energy to cause further displacement when the *first* particle in the event list is followed.

13. Always follow the other particle in the collision, storing the original particle in a list until the particle chosen has insufficient energy to cause further displacement when the *last* particle in the event list is followed.

The procedure used will influence the speed of execution and the memory requirements of the program. Those algorithms which require searching and sorting the event list can be quite slow, especially when the list becomes long, sometimes in excess of 10 000 entries. On older computers, which had only limited amounts of memory available, the size of the event list was a severe restriction on the ability to execute the program at all. In general, collision cascades have a large number of low-energy events, which will terminate quickly without creating further events, so that it is more economical on computer memory always to follow the lowest energy particle in the cascade. Searching and ordering of a list of events, however, will slow the calculation but could achieve a more efficient use of computer memory. If computer memory is a problem then a better solution might be to use an algorithm that chooses the slowest particle most frequently. For cascades produced by high-energy incident ions, because of the small collision cross-sections and hence low probability of large energy transfer, the slowest particle is more frequently the struck particle in the collision. Using algorithms like 12 and 13 above will save the time of sorting and searching the event list and will form only a relatively short list.

7.4.2 Event ordering

Often many events occur simultaneously and where the events overlap can be important in calculating the exact behaviour. The ordering of events can be important. Figure 7.4 shows a simple collision sequence calculated in a time-step regime using two different event store algorithms. In the time-step model the primary particle collides with a target atom, pushing it into a second target particle. If one adopts a fastest recoil first algorithm, similar

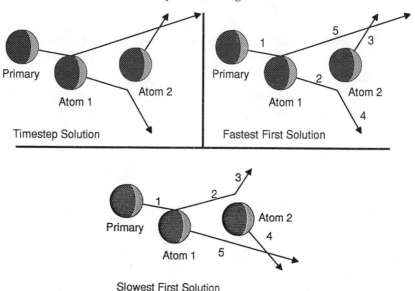

Fig. 7.4 A diagram illustrating that event ordering can affect the particle trajectories.

behaviour is observed to that of the time-step model. If, however, one adopts a slowest-first-algorithm, then the primary particle interacts with the second target atom before the recoil does and so alters the whole behaviour of the ensuing cascade. Thus, depending upon the event store algorithm chosen, one can calculate different paths for the recoiling particles! However, there are many quantities that are not known or calculated with precision. Not only is it not possible to know the exact impact site of a primary particle, but also the target itself is usually in thermal motion and the exact positions of target particles are only determined from a probability distribution. The slowing down of an energetic particle is also determined by its energy loss to the electronic system and this is also dependent upon the distribution of electrons in the target – which is also a time- and space-varying distribution. Provided that the 'average' behaviour of the cascade is unaffected and path errors introduced by the various algorithms are not systematic then the model can still be capable of predicting the average behaviour of an energetic particle–solid interaction.

7.4.3 Non-linear events

If two recoil trajectories overlap in time and space this is often termed a 'non-linear' event, which cannot be described by the linear Boltzmann transport theory. However, as can be seen in Figure 7.5, it is quite easy to

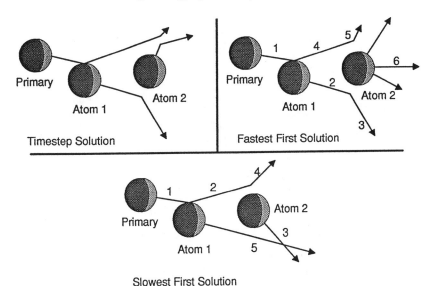

Fig. 7.5 The order of collisions in non-linear situations.

envisage a system in which such non-linear events can take place. How do the algorithms cope with this situation? It can be seen in Figure 7.5 that the exact trajectories of the recoils cannot usually be correctly simulated. One important feature is shown diagramatically for the fastest first algorithm. In this case the recoil labelled 'atom 2' has three possible resultant trajectories depending upon how the algorithm treats multiple impacts on the same particle. In some calculations this will result in two separate recoils being generated from the same place, artificially creating two vacancies there! Care must be taken not to allow this to happen when constructing such a program because this will create an effectively denser target than is realistic, i.e. a systematic error. In other algorithms, the two collisions involving atom 2 are summed vectorially to give a resultant direction of motion. Others just take the first event that arrives. All event store algorithms have a problem when trying to simulate non-linear cascades. It is quite difficult to be certain that systematic errors are not introduced. However, cascades are inherently chaotic. Very small changes in initial conditions will often lead to very large changes in the final cascade. In MD simulations, merely running the same code on two different machines can cause cascade differences. Because these atomic systems are so chaotic, the exact paths of the recoils are less important than the average, effective mass and energy transport of the system as a whole.

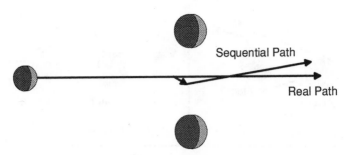

Fig. 7.6 The error in treating a multiple collision as the sum of two binary events.

7.4.4 *Multiple collisions*

A problem arises when a moving particle passes between the centre of two, or more, stationary particles – see Figure 7.6. If we treat this situation as the result of two separate binary collisions then the algorithm must choose one collision as occurring first. Then the initially moving particle has a collision with the second particle and a new scattering angle is deduced. However, this second collision will be based on a reduced energy and maybe even a different collision point; instead of the particle passing through the centre of the pair of atoms without a change of direction, it will have an incorrect energy and direction of motion. If the original target material is amorphous then this may not change the overall statistics. However, if the target is ordered then the incorrect behaviour can alter the channelling behaviour of particles.

A simple way of dealing with this is to calculate such collisions simultaneously and sum the results vectorially. This will give the correct solution for a truly simultaneous collision. However, just how simultaneous is a multiple collision? Clearly, if we slightly move one of the stationary particles then we will still need to calculate the path as a simultaneous collision and not as a succession of separate collisions. It is necessary, therefore, to define a cut-off point at which these should be treated as separated, but not isolated, collisions – see Figure 7.7.

7.4.5 *Incident ion distributions*

In many situations we are only interested in the spatial rest distribution of the energetic ion species in the target matrix, particularly when modelling ion implantation processes. This is described in great detail in Chapter 6. A typical process will involve the implantation of many ions, typically after a dose of more than 10^{14} cm^{-2} onto the surface. Each ion will arrive at a random position on the initial target surface. From there it will undergo a series of collisions both with the atoms and with the electrons of the target

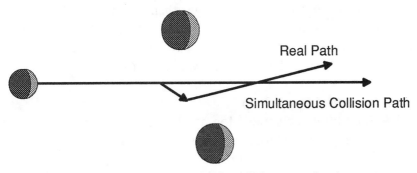

Fig. 7.7 The error in treating a multiple collision as a simultaneous event.

matrix until the particle has insufficient kinetic energy to move further. The stationary particles can be described in terms of a probability distribution as a function of depth below the initial surface. The exact shape of this distribution is determined by the set of all possible collision histories.

If the implantation dose is low enough, each particle can be treated as if it were implanted in isolation and no account need be taken of previously implanted particles. If we assume that the rest distribution of the implanted particles is Gaussian with mean depth, $\bar{x}$, and standard deviation, σ, then the peak concentration N_{max} will be given by

$$N_{max} = \text{Dose}/((2\pi)^{\frac{1}{2}}\sigma).$$

Thus previously implanted particles only start influencing the stopping of new ions when the implantation dose has reached a level such that the maximum concentration is a substantial fraction of the target density, about one atomic per cent. That is

$$\text{Dose} = 0.01 N \sigma (2\pi)^{\frac{1}{2}}$$

where N is the atomic density of the target. For a 50 keV As implantation into Si this will give a dose of about 10^{13} ions cm^{-2} and for higher energies and lower mass particles this value of dose will increase.

If the initial implantation is into a single crystal then the inherent ordering of the target can change the implantation profile substantially due to channelling effects. In this case, previously implanted ions will have caused substantial displacements in the original target and the crystallinity will be gradually degraded, so that ions arriving later will no longer be implanted into a single crystal target but rather into an increasingly damaged material. The exact dose at which this behaviour may start to affect the simulated profile will depend on the rate at which the damage is created and on the rate at which it may be subsequently annealed.

If the implantation dose is low enough, then we only need consider the collisions that the ion has with the target and we do not need to calculate the further development of the recoiled particles. Thus the procedure for calculating an implant profile is as follows:

repeat

       Pick Random Impact Point

       while Ion Still in Target **and** Ion has Energy to Move

            Move IonMeanAtomicSpacing along Path

            Make Random Collision

            Calculate new Ion_path and Ion_Energy

       wend Done Enough Trajectories

until Done Enough Trajectories

7.4.6 *Full cascade models*

In many cases we are also interested in the recoiling target atoms, for example, to calculate implantation damage profiles or to simulate the effects of sputtering. In these cases the simulation needs to follow both the ion and the recoils. This is done using an event store logic and proceeds in the following way:

repeat

  Pick Random Impact Point

  while Ion Still in Target **and** Ion has Energy to Move

       Move Ion MeanAtomicSpacing along Path

       Make Random Collision

       Calculate new Ion_path and Ion_Energy

       Calculate Recoil_path and Recoil_Energy

       while Recoils Exist

            if Recoil Still in Target **and** Recoil has Energy to Move

                 Move Recoil MeanAtomicSpacing along Path

                 Make Random Collision

                 Calculate New_Recoil_Path & Energy

                 Calculate Next_Recoil_Path & Energy

                 Store Recoil Information

                 Set Recoil to Next_Recoil

          else

                 Restore old Recoil Information

          endif

       wend

  wend

until Done Enough Trajectories

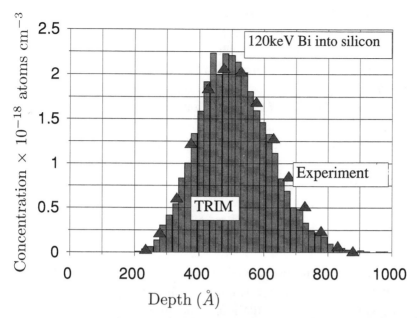

Fig. 7.8 A comparison of a 120 keV Bi implantation profile in an amorphous silicon target from TRIM (Ziegler *et al.*, 1985) and experimental (O'Connor, 1982) results.

7.4.7 *Random materials – Monte Carlo models*

Monte Carlo models are described in detail in Section 6.4. They are an excellent tool in modelling ion implantation profiles. We refer the reader to that section but pass comment here on the implementation of inelastic energy loss within an MC simulation.

During the flight between collisions it is assumed that energy lost by the moving particles, by collisions with electrons, which can be considered to act as a friction-like force over the path of the particle (see Chapter 4). The direction of motion during the flight between collisions with target atoms is unperturbed. The post-flight energy E_l, is given by

$$E_l = E' - T_l,$$

where E' is the energy of the particle after its last collision and T_l is given by equation (6.1).

For simplicity, it is usually assumed that the particle is moving with constant energy between collisions. However, if the energy loss due to electronic processes for that portion of the particles path is subtracted either before or after the scattering calculation is made, systematic errors can be introduced and it is better to assume that the inelastic loss occurs at the point of collision. Equation (2.20) gives the formula for the scattering angle calculated this way. Behar *et al.* (1991) have shown that better agreement with experiment can be achieved for the range of heavy ions implanted into boron films, using this model of inelastic energy loss.

Despite the inaccuracies inherent in the BC algorithm it is possible to obtain very good agreement with experimental predictions of ion range profiles in random materials. In Figure 7.8 a comparison is made between a high-resolution experimental profile (O'Connor, 1982) and a simple Monte Carlo simulation, TRIM (Ziegler, Biersack and Littmark, 1985), for a 120 keV low-dose Bi implantation into an amorphous Si substrate. The agreement between the two is very good.

7.4.8 Input parameters

In timestep (MD) simulations, for Hamiltonian systems, once the potentials have been defined, then the calculation is in principle deterministic. This is not the case in a BC simulation. The numerical values of calculated results depend crucially on a number of input parameters that the BC approximation uses. A typical set of input parameters for the TRIM program for amorphous materials is given in Table 7.1. Most quantities are pre-defined by the particular experiment to be simulated but the values of E_C, E_D, U_0, U_B and E_F are not always well-known quantities.

In the BC model, the collision dynamics are usually derived from purely repulsive potentials. This means that no matter how small the energy of collision, both the projectile and the struck particle would recoil. In a realistic target, a struck particle will not be able to move unless it has sufficient energy to overcome the potential barrier between atoms. The energy required to form a vacancy in this manner tends to be several tens of electronvolts higher than the thermal creation mechanisms. This is difficult to measure (but can be calculated in MD and is directionally dependent in crystals). This energy barrier is called the 'displacement energy', E_D. In order to approximate the time behaviour, in the BC model, if an energy of less than E_D is transferred to the struck particle during a collision, it is assumed not to move. However, should this threshold be exceeded then the struck particle is allowed to take part in the ensuing cascade but with the lattice binding energy, U_B, subtracted from it. There is a similar parameter E_F that determines when to stop calculating the particle trajectory. It is usual to consider E_F to be species-independent and independent of the damage state of the target. The cohesive energy E_C is the energy required per atom to break a solid into separate isolated atoms. This could be regarded as the energy required to

Table 7.1. *Parameters used in a typical BC program*

Name	Parameter	Brief description
Ion data		
E_0	Initial energy	The initial energy given to each ion
m_1	Ion mass	Implanters normally work as isotope separators, so this is usually taken as the most abundant isotope. Needed for stopping power and scattering calculations
Z_1	Atomic number	Needed for stopping power and scattering calculations
θ	Angle of incidence	
Target data		
m_2	Target mass	In a compound this is a table of masses for each constituent element. The average atomic mass is often used
Z_2	Atomic number	In a compound this is a table as above
Stoichiometry		This is a table describing the relative concentration of the constituent elements of the target
$U_B = 2E_C$ (E_C – cohesive energy)	Lattice binding energy	This value is subtracted from every recoil when it is first created
E_D	Displacement Energy	This is the minimum energy a particle must receive to become a recoil. Note that this is different from U_B. Clearly the displacement energy must be greater than or equal to the lattice binding energy.
U_0	Surface binding energy	This is the minimum energy a particle needs to escape from the surface.
E_F	Cut-off energy	When a particle energy drops below this value it is considered to have stopped.

break every bond in the solid. Since each atom is bonded to another, a bulk atom can be regarded as having a binding energy $U_B = 2E_C$.

The surface is taken as a special case. A reduced displacement energy is used in close proximity to a surface. For a particle to be ejected from the surface out of the target it must break the bonds which hold it to the surface. There is, therefore, a potential barrier to be overcome. This is modelled in the simulations by the quantity U_0, the surface binding energy. Often this quantity is taken as the cohesive energy of the target $E_C = U_B/2$, since a surface atom will have approximately half its neighbours present. However, this is not exactly accurate for two reasons. First, the potentials are not pairwise additive and, even if exactly half the neighbours were present, the surface binding energy $U_0 \neq U_B/2$. Secondly, the surfaces reconstruct and are often different in structure from the bulk-terminated lattice. Table 7.2 illustrates these effects using the Tersoff Si potential and the Ackland Ni potential (see Chapter 3).

The position of the surface is also a quantity that varies between BC programs. In most versions of TRIM, the surface is initialised at a distance $2p_t$ above the topmost atom, where the incoming projectile is started. Here p_t is the maximum impact parameter defined in Chapter 6. In other programs the surface is initialised at the topmost atom.

In the application of BC codes to sputtering the exact values for the choice of the input parameters can crucially affect the calculated sputtering yield, a point to which we will return in Section 7.6.2. The important point in a simulation experiment is therefore often not to obtain exact values but to determine general trends by fixing a given number of the input parameters and looking at the effect of varying others.

7.4.9 Single-crystal materials – deterministic models

When the target material is crystalline then the positions of the target material atoms are defined in space, usually with some 'blurring' because of the displacement due to randomisation of the exact atomic position caused by thermal vibrations. Thermal vibrations are discussed in Section 8.11. These models usually ignore correlated thermal motion but this has been attempted in some BC models of surface scattering. In this situation the effects of multiple collisions cannot be ignored. In crystalline materials the regular array of atoms means that, in various directions, crystallographic channels are created in which a moving particle will experience only distant collisions and can travel long distances through the crystal material without losing much energy. It is important in these situations to calculate correctly the exact path of the moving particle and when the particle leaves or enters these channels.

Thus, to model the behaviour of energetic particle impact on crystalline targets, multiple collisions should be identified. This is performed using the translational symmetry of the crystal. A list is made of the atoms in a small region of the crystal using one of the lattice sites as the origin. The size of the region is largely dependent upon the implementation of the program and can be determined by the trade-off in speeds of searching the list of atoms in the region and of creating a new list and reference point as the particle moves towards the edge of the region. Care must be taken so that the region is large enough to include all potential interactions. Often the region is taken as a unit cell of the target material, including nearest and next nearest neighbour atoms. The procedure for searching the crystal for collisions is shown, diagrammatrically, in Figure 7.9. The perpendicular distance from the initial direction of motion of the moving particle to all the atoms in the region is calculated. Those which are greater than p_t are ignored and assumed to be too distant to cause significant deflection or energy loss. Those atoms within p_t of the flight path are analysed and those nearest the initial position of the moving particle are chosen as the next collision partners. By analysing the distances to the various impact points it is possible to determine whether the forthcoming collision should be treated as a binary event or a multiple collision event. The maximum impact parameter p_t is chosen large enough so that in no part of the crystal is the projectile free of the influence of surrounding atoms, but small enough to avoid too many collisions involving very small energy transfers. p_t is often taken as being the average interatomic spacing, but it is also possible to allow p_t to be energy-dependent such that there is a minimum energy transfer to the collision partner – typically of the order 5 eV (Ziegler, Biersack and Littmark, 1985).

7.5 Dynamic models

7.5.1 Dose effects

So far we have only considered the effects of low-dose implantation, during which each new ion trajectory 'sees' a new target. The effects of previous ion trajectories are assumed to have no effect. If we assume that each trajectory initiates a cascade that sets into motion 10 000 atoms, which would be typical of a 500 keV incident particle in an 'average' solid (i.e. not a condensed gas!) this occurs over a depth of about 5000 Å. If each of these displacements were created uniformly along this depth then approximately five atoms per

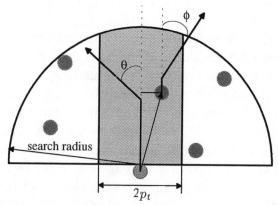

Fig. 7.9 A schematic representation of the search procedure to find the next collision.

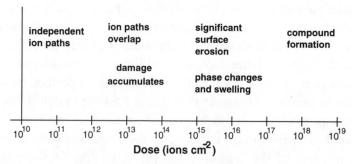

Fig. 7.10 A schematic representation of the changes to a solid caused by increasing ion dose.

monolayer would have moved. In other words assuming that there are approximately 10^{14} atoms cm^{-2} per monolayer then after a dose of about 10^{13} ions cm^{-2}, every particle in the target should have moved once. For doses in the region 10^{11} ions cm^{-2} or lower, there is less than a 1% chance of finding a region of the crystal that has been previously visited by an ion. Above a dose exceeding 10^{12} ions cm^{-2} the changes occurring in the target matrix due to previous trajectories must be considered. In single-crystal targets there is a gradual loss of the initial crystal structure within the dose range 10^{11}–10^{13} ions cm^{-2}.

As the dose increases further the target is sputtered away and the implanted material accumulates, making the target swell. At higher doses still the target material can undergo a phase change as the concentration of implanted material matches or surpasses that of the initial target material. Figure 7.10 illustrates these effects.

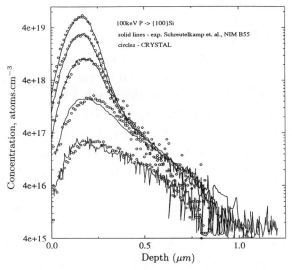

Fig. 7.11 100 keV P→ Si{100} as a function of dose.

If the target is initially a single crystal, it may become gradually amorphous. The degree to which this occurs during the implantation process depends upon a number of parameters. If the target material is initially a stable amorphous material then moving all of the constituent atoms in a random way may have little effect. If the target material can 'self-anneal' then the displaced atoms can move back to lattice sites before the next ion arrives. This often occurs during room temperature implantation of single-crystal metals. If this is not the case then target ordering (crystallinity) can affect the implantation and the damage profiles. Thus, as the target material undergoes a transition from a single-crystal to an amorphous structure, so the implantation profile will change in shape. To model this, one should divide the implantation simulation into a number of separate dose increments, each represented by a number of trajectories (of the order 100) which determines the distribution of disorder to be used for the subsequent dose step. These steps are accumulated until the desired dose is attained. The various implantation parameters can change substantially during this time. Figure 7.11 demonstrates the change in the implantation profile of 100 keV phosphorus implantation into a silicon single crystal as the dose is increased; both experimental (Schreutelkamp *et al.*, 1991) and simulated (Chakarov and Webb, 1994) results are shown.

As the implantation dose increases the bombarded matrix can undergo massive atomic rearrangement via five principal mechanisms.

  (i) Ion implantation, in which the ions themselves are implanted, adding to the total number of particles in the material as a whole.

  (ii) Recoil implantation, in which target material can be driven deeper into the target by direct collision with the high-energy ion beam. This tends to decrease the density of the surface region and increase the density of the deeper regions of the target.

 (iii) Cascade mixing, in which the target material is generally redistributed isotropically, broadening layers and smearing interfaces.

  (iv) Sputtering, in which atoms from the surface region of the target are stripped away from the target, causing the target surface to move inwards.

  (v) Target relaxation, in which the transitory changes in density of the target material are removed and the final positions of the surface and other boundaries are established. For example, ignoring any density changes caused by phase changes, if the sputtering yield is greater than unity then the surface moves inwards. If the sputtering yield is less than unity (more particles implanted than removed) then the target will swell and the surface moves outwards.

Dynamic computer programs such as TRIDYN (Möller and Eckstein, 1984) deal with the effects described above in amorphous materials. The target is divided into a series of layers parallel to the surface. A simulation is performed (typically 100 trajectories) to represent the dose increment during which a perturbed target can be created in which sputtering has taken place from the surface and a certain number of particles has been implanted within the material. Each layer thickness can then be adjusted depending upon how many particles are now contained in each layer, see Figure 7.12. Once this adjustment has been made, this new resultant layer structure is used for the next dose increment. This sequence of events is repeated until the required implantation dose is reached. The book by Eckstein (1991) describes the TRIDYN program in detail.

7.5.2 *Annealing effects*

The simulation of high-dose implantation can be further complicated if the defects created during the implantation are unstable and anneal or migrate during subsequent dose increments. For example, in the case of 80 keV As implantation of Si the models discussed so far can be used to calculate the accumulation of displaced atoms for different initial doses (Budinov and Karpuzov, 1991). This can be compared with experimental measurements

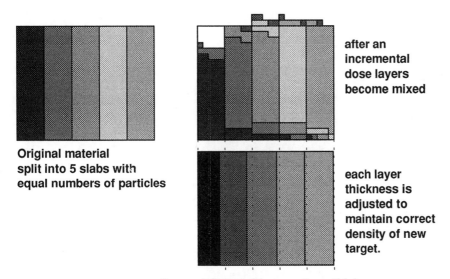

**Original material
split into 5 slabs with
equal numbers of particles**

**after an
incremental
dose layers
become mixed**

**each layer
thickness is
adjusted to
maintain correct
density of new
target.**

Some slabs get thinner, others thicker.

Fig. 7.12 The incremental dose scheme used to model high-dose effects.

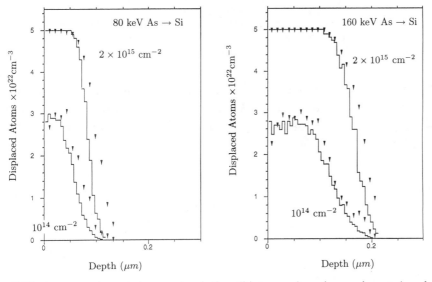

Fig. 7.13 A comparison between simulation (histogram) and experiment (symbols) for As disorder profiles in Si, for an implantation dose of 2×10^{15} ions cm^{-2}: (a) 80 keV As and (b) 160 keV As.

and good agreement (Budinov and Karpuzov, 1991) can be achieved, see Figure 7.13. However, when the same calculation is performed for 80 keV boron and compared with experimental data the comparison is poor, see Figure 7.14.

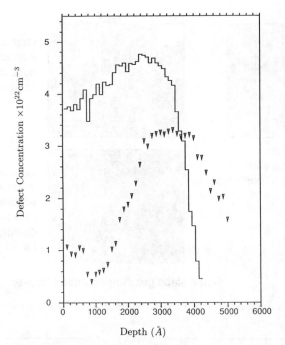

Fig. 7.14 A comparison between simulation (histogram) and experiment (symbols) for 80 keV B disorder profiles in Si. Defect annihilation and migration are not included.

A simple diffusion model to allow the defects to migrate and recombine with each other was employed (Budinov and Karpuzov, 1991) to improve substantially the agreement between the simulation and experiment, see Figure 7.15.

7.6 Applications

7.6.1 Ion scattering spectroscopy

The BC approximation can be used effectively to predict the spectra observed in ion scattering spectroscopy (ISS). It is therefore a useful tool in surface structure investigations of crystalline materials. The simulations are sensitive to lattice vibrations and these should be included in the simulations. The inclusion of lattice vibrations increases the running time of the codes but this is becoming less of a problem as computing time becomes ever cheaper. However, Tromp and Van der Veen (1983), have developed an approach using hitting probability integrals that speeds up the computing time for high-energy scattering in which particles undergo many scattering events.

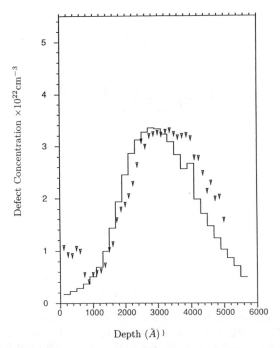

Fig. 7.15 A comparison between simulation (histogram) and experiment (symbols) for 80 keV B disorder profiles in Si. Defect annihilation and migration are included.

Those authors have also developed a sophisticated Monte Carlo code called VEGAS, which is tailored to ISS simulations.

In ISS a beam is projected towards a surface at some angle of incidence and azimuthal orientation of the substrate and the scattered particles are detected only in a relatively small solid angle. The number of ions scattered into this solid angle is small and so it is often the case that many hundreds of thousands of trajectories should be calculated for good statistics. Nonetheless, good agreement between experiment and theory can often be obtained. Figure 7.16 compares the experimental and simulation results (MARLOWE) for 2 keV Ne ions incident at 9.5° on Ir{110}(1 × 1).

In the analysis of impact collision ISS (ICISS) experimental data, the ion detector is located close to the ion source so that particles backscattered through 180° are detected. Thus only those particles which make almost direct hits with the surface layer atoms will be backscattered. It is thus possible to speed up the backscattering calculations by selectively sampling the irreducible symmetry zone of the crystal to include only those points at which backscattering is likely. Backscattering through angles close to 180° is a relatively rare event for an ion and typically we need many thousands of trajectories for good statistics. The authors have written such a program

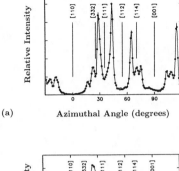

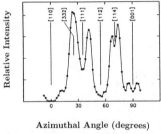

(a)   Azimuthal Angle (degrees)

(b)   Azimuthal Angle (degrees)

Fig. 7.16 (a) The intensity distribution for 2 keV Ne$^+$ ions incident at 80.5° to the normal of the Ir{100} surface as a function of azimuth. The ions are collected after scattering through 40° (courtesy W. Heiland). (b) The MARLOWE simulation.

(Smith and Body, 1994). The program typically takes about 4 min on a 33 MHz 486 PC for 250 000 trajectories. The program preferentially samples impact points in regions where backscattering is more likely to occur.

For scattering of light ions or ions that only undergo a few collisions with the lattice, we need only follow the ion trajectory and not the recoils and so the basis of the algorithm is as follows.

1. Divide the crystal into symmetric cells and treat the first collision as a special case.
2. For an ion having undergone a collision with an atom at a point whose position vector is r, work out a new neighbour list L of n possible collision partners by 2.1 and 2.2.
2.1. Adding to the atom list all points in the unit cell containing r.
2.2. Adding to the atom list all points in the next unit cell which contains $r + tv$ for minimum t, where v is the ion velocity.
3. Choose the next target atom as that with position vector y, where $(y - r) \cdot v = \min \{(x - r) \cdot v : x \in L, (x - r) \cdot v \geq 0, p(x) < p_t\}$. Here p is the impact parameter defined by $p(x) = |(x - r) \times v| / |v|$ and p_t is the maximum value we allow for an interaction to take place.
4. Repeat the process from 2 onwards. The trajectory is terminated if

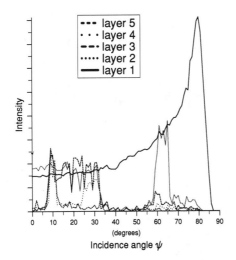

Fig. 7.17 The intensity distribution for 10 keV Ne incident upon Au{100} backscattered through 160°–180° as a function of the incidence angle of the beam. The azimuth is the ⟨100⟩ direction.

the ion passes too deep into the crystal, scatters out, its energy drops below a specified threshold or a maximum number of collisions is exceeded.

The energy spectrum of backscattered ions can be used to determine the target composition and in this case inelastic loss mechanisms must be included in the algorithm. Figure 7.17 shows a typical result from the program, which illustrates the effects of the crystal structure on the backscattering yield for 10 keV Ne^+ ions on Au{100} as a function of incidence angle. The calculations have been carried out for backscattering through angles >160°, as might be obtained in an ICISS instrument. Figure 7.17 shows not only the effects of crystal structure but also the extra information which a simulation can give over an experiment, namely from which layer the ions were scattered.

Calculated scattering patterns can be tested against experimentally measured spectra to assist with the determination of crystal structure. It is now fairly straightforward to predict the scattering pattern from a known crystal structure. The inverse problem is more difficult, i.e. given an experimentally observed scattering spectrum, what is the corresponding crystal structure which produces that spectrum? It is not even clear whether there is a unique solution to this problem.

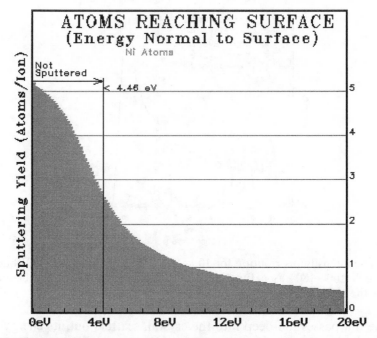

Fig. 7.18 The cumulative energy distribution of particles arriving at an amorphous Ni surface calculated using TRIM95. If $U_0 = 4.46$ eV, the calculated sputtering yield is 2.68 atoms per ion. The backscattering yield is 0.049.

7.6.2 Sputtering

The BC model has been used extensively in the calculation of sputtering yields, see for example the book by Eckstein (1991) but it is also an area in which caution should be used concerning the exact values given by the various programs because these values are crucially dependent on the input parameters and the form of BC model used.

In order to calculate the sputtering yield, it is assumed that the surface binding energy acts as a barrier to the ejecting particles. If this is assumed to be a planar barrier then only the normal component of velocity of the particle is changed and thus the particle is refracted by the barrier. If the particle arrives at the surface with energy E at an angle θ to the normal then, after passing through the barrier, the energy and direction of the particle will be

$$E' = E - U_0,$$

$$\cos \theta' = \left[(E \cos^2 \theta - U_0)/(E - U_0) \right]^{\frac{1}{2}}.$$

An isotropic surface binding model has also been used in which no refraction of the ejected material takes place but the final energy is the same as that

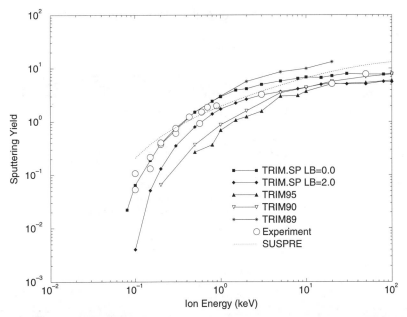

Fig. 7.19 The sputtering yield for Xe ions incident on amorphous Ni as a function of ion energy.

of the planar model. It is generally thought that the planar model gives the best results for metals.

Figure 7.18 shows the cumulative energy distribution of all particles reaching the surface of an amorphous Ni target during 1000 individual trajectories after 2 keV Ar bombardment, calculated using TRIM95[1]. The surface binding energy U_B is marked on Figure 7.18. It can be seen that small changes to this value can cause substantial changes in the number of ejected particles.

In Figure 7.19 the sputtering yield for Xe ions impacting on amorphous Ni is plotted as a function of energy. The open circles show a range of experimental measurements (Anderson and Bay, 1981). Also plotted are the results from the SUSPRE[2] program and five simulations involving different versions of TRIM. According to the software documentation, there have been no major differences in the treatment of the surface. All the simulations were performed using $U_0 = 4.46$ eV and $E_D = E_F = U_B = 0$. The lattice binding energy was varied in the TRIM.SP (Biersack and Eckstein, 1984) program to demonstrate the effect that this parameter had on the sputtering yield. There are thus substantial differences between the various versions of

---

[1] The TRIM85-TRIM95 program is available from J. Ziegler, IBM, Yorktown Heights, NY, USA.
[2] SUSPRE is a program based on energy deposition in the surface and is available from R. P. Webb, University of Surrey, UK.

Binary collision algorithms

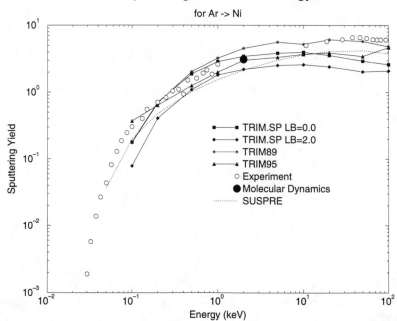

Fig. 7.20 The sputtering yield for Ar ions incident on amorphous Ni as a function of ion energy.

Table 7.2. *Values for the potential energies of particles in the bulk*
(E_C) and the surface (E_S) for Si and Ni calculated using the Tersoff
and Ackland potentials described in Chapter 3

Element	E_C (eV/atom)	E_S (eV/atom) {100}	{110}	{111}	{100}(2 × 1)
Si	4.63	2.50	2.50	3.60	3.49
Ni	4.48	3.88			

TRIM. The main differences arise because of the way in which electronic energy loss is included. In TRIM89, this loss is subtracted *after* the collision. In later versions it is subtracted *before* the collision. A similar picture is shown in Figure 7.20 for the case of Ar incident on Ni. Here we also plot the sputtering yield using MD for the {100} surface (with electronic energy loss included). Figure 8.20(d) shows the state of a typical cascade after the crystal has annealed, using MD.

All the BC calculations were carried out using $U_B = 0$. It was argued by Biersack and Eckstein (1984) that the value of U_B does not affect the sputtering yield substantially. However, this is clearly an over-simplification and can affect particles close to the surface in the same way as does U_0. Thus, the sensitivity of the sputtering yields calculated by TRIM both to the BC model and to the parameters used in the model highlights the difficulty in using the BC approximation for sputtering simulations.

Many BC codes have had great success in modelling ion implantation distributions and, whereas changes to the values of the energy input parameters and the exact nature of the surface can make little difference to the implantation profiles, they can lead to substantial changes in the values of absolute sputtering yields. This demonstrates that such model extensions must be used with extreme care.

7.6.3 *Trajectories and displaced particles*

Figure 7.21 gives the results of a typical trajectory of a 2 keV Ar ion impacting on an amorphous Ni target, using TRIM95. The figure shows both the ion trajectory and a two-dimensional projection of the places at which Ni atoms have been struck with sufficient energy to move them from their initial sites, in this case over 40. For this trajectory two Ni atoms are ejected from the surface, and the *Ar* atom comes to rest at a depth of 30 Å.

Figure 7.21 should be compared with Figure 8.20(c), which calculates a similar trajectory for normal incidence on the {100} face of Ni using MD. The MD trajectory is run for over 7 ps so that annealing has taken place. Two particles are sputtered. Four adatoms form and there are four surface vacancies. The target recrystallises after the initial damage so that there are fewer sub-surface defects than in the BC simulation. Most of them are in the form of dumb-bell interstitials but the ion which comes to rest at a depth of 215 Å has vacancies surrounding it, owing to the purely repulsive nature of the Ar/Ni interaction potential.

Thus, in order to obtain closer agreement with MD the BC model must include the effects of annealing. The number of *initial* displaced particles is not enough to produce a good description of cascade damage and the annealing models described earlier should be used.

7.7 Conclusions

It is possible with care to predict many features of energetic particle–solid interactions by using a much simplified model. At the present time it

Depth (Angstroms)

0.0   20.0   40.0

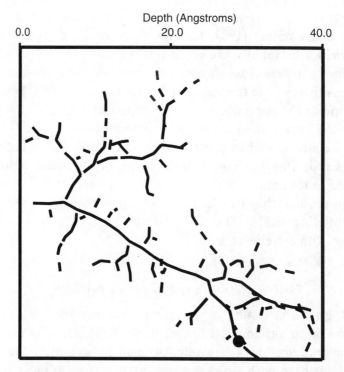

Fig. 7.21 The ion trajectory and initially displaced atoms for 2 keV Ar incident upon Ni, calculated using TRIMSP, with $U_B = 2$ eV, $U_0 = 10$ eV, $E_F = 10$ eV and $E_D = 15$ eV. The unbroken line is the ion path; the broken lines are the recoil trajectories. The circle is the final resting place of the ion.

is still neither possible to perform a full molecular dynamics calculation over the time scales needed to simulate the effects of high- or medium-dose implantation, nor even to observe the effects of defect migration and recombination for the high energies and large doses used in practice. The simulation of the relatively rare event of ion scattering from surfaces can also be very expensive in computing time. Until such time that this is possible, simplifying assumptions, such as those outlined in this chapter, will be needed to enable the prediction of the behaviour of the interactions in a real-time environment.

References

Anderson, H. H. and Bay, H. L. (1981) *Sputtering by Particle Bombardment* Vol. 1 Ed. R. Behrisch, Springer-Verlag, Berlin, 175.

Behar, M., Grande, P. L., Wagener de Oliviera, R. and Biersack, J. P. (1991). *Nucl. Instrum. Meth.* B **59/60** 1.

Biersack, J. P. and Eckstein, W. (1984). *Appl. Phys.* A **34** 73.

Budinov, H. I and Karpuzov, D. S. (1991). *Nucl. Instrum. Meth.* B **59/60** 1041.

Chakarov, I. R. and Webb, R. P. (1994). *Rad. Eff. Defects Solids* **130** 447.

Chakarov, I. R., Webb, R. P., Smith, R. and Beardmore, K. (1995). *Nucl. Instrum. Meth.* B **102** 145.

Eckstein, W. (1991). *Computer Simulation of Ion Solid Interactions*, Springer-Verlag, Berlin.

Möller, W. and Eckstein, W. (1984). *Nucl. Instrum. Meth.* B **2** 814.

O'Connor, D. J. (1982). *Nucl. Instrum. Meth.* **196** 493.

Robinson, M. T. (1989). *Phys. Rev.* B **40** 10717.

Robinson, M. T. and Torrens, I. M. (1974). *Phys. Rev.* B **9** 5008.

Schreutelkamp, R. J., Raineri, V., Saris, F. W., Kaim, R. E. and Westendorp, J. F. M. (1991). *Nucl. Instrum. Meth.* B **55** 615.

Smith, R. and Body G. (1994). *Vacuum* **45** 733.

Tromp, R. M. and Van der Veen, J. F. (1983). *Surf. Sci.* **133** 159.

Vichev, R. G. and Karpuzov, D. (1993). *Nucl. Instrum. Meth.* B **83** 345.

Webb, R. P., Smith, R., Dawnkaski, E., Garrison, B. and Winograd, N. (1993). *Int. Video J. Eng. Res.* **3** 163.

Ziegler, J. F., Biersack, J. P. and Littmark, U. (1985). *Stopping Power and Ranges of Ions in Matter*, Vol. 1 Ed. J. F. Ziegler, Pergamon, New York.

8

Molecular dynamics

8.1 Equations of motion

Molecular dynamics (MD) can be described as the computation of the motion of systems of particles from a knowledge of the interaction forces between the particles. Thus, if we denote the position, velocity and acceleration of the ith particle of a system of N particles by r^i, v^i and a^i respectively, the dynamics of the system can be determined by solving Newton's laws of motion for each particle:

$$\dot{r}^i = v^i,$$
$$\dot{v}^i = a^i = F^i\left(r^1, \ldots r^N, v^1, \ldots v^N, t\right)/m^i, \tag{8.1}$$

where m^i is the mass of the ith particle which is subject to a force, F^i, assumed to depend only on the positions, velocities and type of the N particles in the system and the time t.

Many problems involve non-dissipative systems in which F^i depends only on the position vectors $r^j, j = 1, \ldots N$ and in addition is derivable from an interatomic potential function, $F^i = -\nabla^i V$. In some cases inelastic losses such as electronic energy loss can be modelled by the assumption that the dissipative forces are velocity-dependent. However, for a non-dissipative system of Hamiltonian form, there are certain properties that can be utilised in the construction of the integration algorithm. It will be our intention in this chapter to describe numerical algorithms that make use of some of the conservation properties of Hamiltonian systems.

The trajectories of particles for Hamiltonian systems can be determined from Hamilton's principle: 'The motion of a system from times t_1 to t_2 is such that the line integral of the Lagrangian $\mathscr{L}$ is an extremum', namely

$$\delta \int_{t_1}^{t_2} \mathscr{L} \, dt = 0. \tag{8.2}$$

234

The Lagrangian $\mathcal{L}$ is the difference between the kinetic and potential energies of the system and is given for a two-body interaction in equation (2.26). The Hamiltonian $\mathcal{H}$ is defined by

$$\mathcal{H}(\boldsymbol{q},\boldsymbol{p},t) = \sum_{i=1}^{3N} \dot{q}^i p^i - \mathcal{L}(\boldsymbol{q},\dot{\boldsymbol{q}},t), \tag{8.3}$$

where $\boldsymbol{q}$ are the generalised co-ordinates $\boldsymbol{q} = \left(\boldsymbol{r}^1,\dots\boldsymbol{r}^N\right)$ and the momenta $\boldsymbol{p} = \left(m^1\boldsymbol{v}^1,\dots m^N\boldsymbol{v}^N\right)$. The generalised momentum p^i is related to q^i thus:

$$\boldsymbol{q} = \left(\boldsymbol{r}^1\dots\boldsymbol{r}^N\right) = \left(q^1,\dots q^{3N}\right),$$
$$\boldsymbol{p} = \left(m^1\boldsymbol{v}^1\dots m^N\boldsymbol{v}^N\right) = \left(p^1,\dots p^{3N}\right)$$

with

$$p^i = \frac{\partial \mathcal{L}}{\partial \dot{q}^i}.$$

Hamilton's equations equivalent to (6.1) can be written

$$\dot{q}^i = \frac{\partial H}{\partial p^i}; \quad -\dot{p}^i = \frac{\partial H}{\partial q^i}. \tag{8.4}$$

The $6N$-dimensional space $(\boldsymbol{q},\boldsymbol{p})$ is called the phase space of the system. Any evolution of the system from initial co-ordinates $(\boldsymbol{q}(0),\boldsymbol{p}(0))$ to $(\boldsymbol{q}(t),\boldsymbol{p}(t))$ at time t is a canonical transformation (Sanz–Serna, 1992) and by Liouville's theorem preserves volume in phase space. The flow also preserves the Poincaré invariants and the sum of areas of projections of any two-dimensional surface S in phase space on to the q^i–p^i planes.

8.2 Numerical integration algorithms

8.2.1 Hamiltonian systems

A numerical algorithm that is canonical (to the order of accuracy of the method) is termed a symplectic algorithm and it has been shown that such algorithms can have better energy conservation properties after integrating for large numbers of timesteps (Channel and Scovel, 1990). A review of these symplectic numerical methods has been given by Sanz-Serna (1992).

In this section we will describe our own preferences of numerical integration method, distinguishing between the algorithms that are symplectic and those that are not. For a non-specialist in numerical methods, the choice of integration algorithm can be quite difficult. Hundreds of methods for integration exist in the literature and sometimes books on MD can just give a list of the most popular rather than a useful discussion on which to use. There are only two requirements that are of any real concern in a simulation code, namely, speed and accuracy. Storage of data in computer memory is no longer an important consideration for most modern high-speed computers running an MD code. It would seem at first sight that one should use a high-order algorithm, which means that longer timesteps can be used for a given accuracy but, also, one that minimises the number of force evaluations per timestep, since evaluating the forces is often very time-consuming. This means that multistep methods are to be preferred over the Runge–Kutta methods since they require fewer force evaluations per timestep for the same truncation error. In practise, the situation is not quite so straightforward, for two reasons. The timestep is generally not fixed when simulating energetic collision cascades but is calculated by the program. If a particle is travelling fast or subjected to a large force, then a smaller timestep is required for a given order of accuracy. The program usually takes account of this. Multi-step methods are best used with fixed timesteps. Secondly, MD simulations may also use neighbour lists, often with a cut-off radius r_c, such that the contribution to the force F^i on the ith particle contains no dependence on r^j if $|r^i - r^j| > r_c$. The multistep methods can contain built-in redundancy if we are storing lots of forces from previous timesteps that contribute little to the dynamics.

The most robust low-order algorithm to have been tried and tested over many years is the Verlet algorithm (Verlet, 1967). This is derivable from a Taylor series expansion for r about the nth timestep:

$$r_{n+1} = r_n + v_n \Delta t_n + \tfrac{1}{2} a_n \Delta t_n^2 + O(\Delta t_n)^3. \tag{8.5}$$

In this equation we have omitted the i superscript for clarity and the subscript denotes the timestep number at which the physical quantity is evaluated. This equation is second order in Δt_n and the acceleration a_n is related to the force F_n by Newton's second law,

$$F_n = m a_n.$$

For the case in which the force on each particle depends only on position, it is convenient to evaluate the forces at the $(n + 1)$th timestep after updating r_n to r_{n+1}, so that a forward difference formula for the derivative of the acceleration

$$a'_n = (a_{n+1} - a_n)/\Delta t_n + O(\Delta t_n)$$

can be substituted into the Taylor series for v_{n+1} to give a formula for v_{n+1}

which is also of second order

$$v_{n+1} = v_n + \tfrac{1}{2}(a_{n+1} + a_n)\,\Delta t_n + O(\Delta t_n)^3. \tag{8.6}$$

Equations (8.5) and (8.6) form the Verlet algorithm. In the case of Hamiltonian systems, the Verlet algorithm turns out to be symplectic. However, it is only of order 2, so small timesteps have to be used in the simulations. Nonetheless, this algorithm is highly recommended for use in MD simulations. The algorithm is similar to another well-known algorithm, the leap frog algorithm, but is more convenient to use since, in the leap frog algorithm, the velocities and displacement are evaluated at times that differ by half a timestep. This makes book-keeping checks on energy conservation less efficient.

The same idea can be used to produce higher order methods that require only one force evaluation per timestep. The values of a_{n-1} can be stored and then a third-order, two-step method can be derived by using backward differences to approximate a_n' in the Taylor series expansion for r and then central differences for a_n' and a_n'' based on the newly calculated value of r_{n+1} in the Taylor series expansion of v.

This leads to a third-order, two-step algorithm (Smith and Harrison, 1989):

$$r_{n+1} = r_n + v_n\,\Delta t_n + [(3+R)a_n - Ra_{n-1}]\,\Delta t_n^2/6, \tag{8.7}$$

$$v_{n+1} = v_n + \left(\frac{3+2R}{1+R}a_{n+1} + (3+R)a_n - \frac{R^2}{1+R}a_{n-1}\right)\frac{\Delta t_n}{6}. \tag{8.8}$$

Here R is the timestep ratio $\Delta t_n/\Delta t_{n-1}$. Ways of determining Δt_n will be described shortly.

The algorithm described by equations (8.7) and (8.8) is equivalent to an Adams–Bashforth method for r_{n+1} and an Adams–Moulton method for v_{n+1}. The forces are again evaluated after the calculation of r_{n+1}, before determining v_{n+1}. The benefit over the Verlet algorithm is that longer timesteps can be used because it is of third order. The disadvantages are that it is not symplectic and requires extra storage for the values of a_{n-1}. However, on most modern high-speed computers, storage is not a problem and the limiting factor in MD simulations is the speed of the machine, rather than the storage of a_{n-1}. Storage can sometimes be a problem for systems with large numbers of particles but it is the size of the neighbour lists rather than the integration algorithm, which is the determining factor. The method is also not self-starting because a_{-1} is not defined. The Verlet algorithm might therefore be used over the first timestep. In principle, we can extend the multistep method to any order by storing the acceleration and its derivatives from previous timesteps. The multistep methods therefore have the great advantage of speed over the Runge–Kutta methods for the same order of accuracy. A non-Hamiltonian-specific third order explicit Runge–Kutta method would generally require three force evaluations per timestep, a fourth-order method would require four and a fifth-order method six. Implicit methods can appear to require fewer force evaluations but then iterative methods have to be used to solve the non-linear equations involved and, although implicit methods are robust and stable, they are very expensive in computing time.

Hamiltonian-specific Runge–Kutta algorithms in which $\boldsymbol{a} = \boldsymbol{a}(\boldsymbol{r})$ can be expressed in the following form:

$$\boldsymbol{k}_i = \boldsymbol{r}_n + c_i \boldsymbol{v}_n \, \Delta t + \sum_{j=1}^{i-1} a_{ij} \boldsymbol{a}(\boldsymbol{k}_j) \, \Delta t^2,$$

$$\boldsymbol{r}_{n+1} = \boldsymbol{r}_n + \boldsymbol{v}_n \, \Delta t + \sum_{i=1}^{s} b_i \boldsymbol{a}(\boldsymbol{k}_i) \, \Delta t^2,$$

$$\boldsymbol{v}_{n+1} = \boldsymbol{v}_n + \sum_{i=1}^{s} B_i \boldsymbol{a}(\boldsymbol{k}_i) \, \Delta t,$$

where s is the order of the method. The constants a_{ij}, b_i, B_i and c_i are defined by the method and the method is explicit if the matrix (a_{ij}) is strictly lower triangular. Explicit, symplectic Runge–Kutta methods do exist. The one- and two-stage explicit symplectic Runge–Kutta methods are both equivalent to the Verlet algorithm. The only explicit two-stage method of order 3 has imaginary coefficients and is therefore difficult to work with. There is a three-stage, fourth-order method due to Forest and Ruth (1990). The coefficients are given by

$$c_1 = \frac{1}{2} - \gamma, \qquad c_2 = \frac{1}{2}, \qquad c_3 = \frac{1}{2} + \gamma,$$

$$B_1 = \frac{1}{24\gamma^2}, \qquad B_2 = 1 - \frac{1}{12\gamma^2}, \qquad B_3 = \frac{1}{24\gamma^2},$$

$$b_i = B_i(1 - c_i), \qquad i = 1, \ldots 3, \qquad a_{ij} = B_j(c_i - c_j) \, i, j = 1, \ldots 3,$$

where γ is the real zero of $48x^3 - 24x^2 + 1 = 0$:

$$\gamma = -0.175\ 603\ 595\ 798\ 288. \tag{8.9}$$

The three algorithms defined by equations (8.6)–(8.9) should be sufficient for any system to be studied and in the author's view represent the 'best' that exist at the present time of writing. Methods other than these, such

Table 8.1. *Properties of the algorithms*

Algorithm	Order	Number of force evaluations per timestep	symplectic
Verlet	2	1	yes
Two-step	3	1	no
Forest and Ruth	4	3	yes

as Gear's method (Allen and Tildersley, 1987), are also in common use. Higher order methods could be used but there seems little to be gained by using algorithms of order greater than 4. Some of the empirical potentials described in Chapter 3 are discontinuous in their second derivative. For such potentials, the Verlet method conserves energy better than do any of the higher order methods. Table 8.1 summarises some of the important properties of the algorithms described above.

8.2.2 *Constraint dynamics*

For high-energy collisions with crystal surfaces, the algorithms defined in Section 8.2.1 can be used directly because the forces acting on the particles are derivable from a potential function that depends only on the position of the particles. In polyatomic systems, at lower energies, the interatomic bonds may stretch, bend and twist. Often any motion that takes place can do so subject only to constraints of some kind. For example, we might insist that the bond lengths remain fixed in a polyamatic molecule. If the interatomic potentials are good enough, then bond-length constraints or constraints limiting the degrees of freedom of a molecule should not be necessary. Good interatomic potentials should account for bond rigidity. However, much constrained computational chemistry is still carried out and so we mention briefly an efficient algorithm for dealing with constraints that works well when the interatomic forces are derivable from pairwise additive potentials.

A well-tried method, the RATTLE algorithm (Anderson, 1983) for solving the equations of motion subject to holonomic constraints, is to solve the equations of motion at each timestep in the absence of the constraint forces and then iterate to determine their magnitude and connect the atomic positions. Here we describe how the Verlet algorithm may be efficiently adapted for the case of constrained motion.

For the case of constrained motion in which the ℓ constraints are of the form $c_k(\boldsymbol{q},t) = 0$, $k = 1,\ldots\ell$, the method of Lagrange multipliers can be used together with Hamilton's principle (8.2) to produce modified equations of motion. These result in an extra force term

$$\boldsymbol{g}^i = \sum_{k=1}^{\ell} \lambda_{ik}\nabla_i c_k$$

on the ith particle and then

$$m^i\boldsymbol{a}^i = \boldsymbol{f}^i + \boldsymbol{g}^i, \tag{8.10}$$

where $\boldsymbol{f}^i$ is the force derivable from the interatomic potential. For a system of bond-length constraints of the form

$$d_{ij}^2 - r_{ij}^2 = 0, \tag{8.11}$$

where d_{ij} is the fixed distance between the i and j particles and $r_{ij} = |\boldsymbol{r}_{ij}| = |\boldsymbol{r}^i - \boldsymbol{r}^j|$, we can associate a Lagrange multiplier $\frac{1}{2}\lambda_{ij}$ with each constraint, giving

$$\boldsymbol{g}^i = \sum_k \lambda_{ik}\boldsymbol{r}_{ik}. \tag{8.12}$$

For pair potentials, the parameters λ_{ik} satisfy $\lambda_{ik} = -\lambda_{ki}$ and, if there are ℓ distinct constraints, then there are ℓ distinct λ_{ik} terms. These λ_{ik} terms have first to be determined before $\boldsymbol{r}^i(t + \Delta t)$ can be calculated. A possible algorithm which is of second order is as follows.

Obtain a first guess λ_{ij} at the nth timestep λ_{ij}^{A}. The values from the previous timestep might be a possible first approximation:

$$\boldsymbol{g}^i = \sum_k \left(\lambda_{ik}^{A} + \delta\lambda_{ik}\right)\boldsymbol{r}_{ik}. \tag{8.13}$$

From equation (8.5)

$$\boldsymbol{r}^i(t + \Delta t) = \boldsymbol{r}^i(t) + \boldsymbol{v}^i(t)\,\Delta t + \frac{1}{2m^i}\left[\boldsymbol{f}^i(t) + \boldsymbol{g}^i(t)\right]\Delta t^2 + O(\Delta t)^3 \tag{8.14}$$

or

$$\boldsymbol{r}^i(t + \Delta t) = \boldsymbol{r}_{A}^i(t + \Delta t) + \frac{1}{2m_i}\sum_k \delta\lambda_{ik}\,\boldsymbol{r}_{ik}\,\Delta t^2,$$

where

$$\boldsymbol{r}_{A}^i(t + \Delta t) = \boldsymbol{r}^i(t) + \boldsymbol{v}^i\,\Delta t + \frac{1}{2m^i}\left(\boldsymbol{f}^i(t) + \sum_k \lambda_{ik}^{A}\boldsymbol{r}_{ik}\right)\Delta t^2. \tag{8.15}$$

We can obtain $\delta\lambda_{ij}$ by an iterative scheme. First assume that, for any pair of atoms i and j, all the λ terms are exact except λ_{ij} so that

$$r^i(t + \Delta t) = r^i_A(t + \Delta t) \frac{1}{2m^i} \delta\lambda_{ij} r_{ij}(t) \Delta t^2,$$

$$r^j(t + \Delta t) = r^j_A(t + \Delta t) - \frac{1}{2m^j} \delta\lambda_{ij} r_{ij}(t) \Delta t^2. \tag{8.16}$$

From equations (8.11) and (8.16)

$$d^2_{ij} - \left| r^i_A(t + \Delta t) - r^j_A(t + \Delta t) \right|^2 = \Delta t^2 \left(\frac{1}{m^i} + \frac{1}{m^j} \right) \delta\lambda_{ij}$$

$$\times r_{ij}(t) \cdot \left[r^i_A(t + \Delta t) - r^j_A(t + \Delta t) \right] + O(\Delta t^4). \tag{8.17}$$

This is solved for $\delta\lambda_{ij}$ and the result used to improve $r^i_A(t+\delta t)$ and $r^j_A(t+\delta t)$. The calculation is repeated pairwise until all constraints have been exhausted. The procedure is iterated until all the bonds satisfy the constraint equations. This algorithm, which is a minor modification of the SHAKE (Ryckaert *et al.*, 1977) and RATTLE algorithms, requires only one force evaluation per timestep.

8.2.3 Numerical integration algorithms for non-Hamiltonian systems

When an MD simulation is required for systems involving inelastic losses, the numerical algorithms given before can no longer be applied, because a_n may depend on v^i and r^i $(i = 1,\dots N)$ and the system is not Hamiltonian. An algorithm similar to the Verlet algorithm does exist for these cases and can be derived in place of equation (8.6) by noting that

$$a'_n = (a_n - a_{n-1}) / \Delta t_n + O(\Delta t_n)$$

and substituting into the Taylor series expansion of v

$$v_{n+1} = v_n + a_n \Delta t_n + \frac{1}{2} a'_n \Delta t_n^2 + O(\Delta t_n^3).$$

We obtain

$$v_{n+1} = v_n + \frac{1}{2} (3a_n - a_{n-1}) \Delta t_n + O(\Delta t_n^3). \tag{8.18}$$

in place of (8.6). In this case the forces at the $(n-1)$th timestep must now be stored (Smith and Harrison, 1989).

The algorithm is now slightly more complicated to apply because, as with all multistep methods, the first timestep must be treated as a special case.

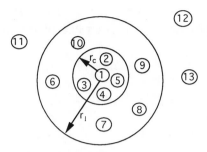

Fig. 8.1 The potential cut-off sphere and the neighbour list sphere around atom 1. Atoms 2–5 contribute to the force on atom 1. Atoms 2–10 are on the neighbour list of atom 1.

Similarly for the higher order method, equation (8.8) must be replaced by the Adams–Bashforth, rather than by the Adams–Moulton, formula.

In the case of constant timesteps this becomes

$$v_{n+1} = v_n + \frac{1}{12} (23a_n - 16a_{n-1} + 5a_{n-2}) \Delta t_n \tag{8.19}$$

and extra storage is required for the forces at the $(n-2)$th timestep. If a variable timestep is employed, then the equation involves the ratios $\Delta t_n / \Delta t_{n-1}$ and $\Delta t_n / \Delta t_{n-2}$.

Other algorithms, such as Runge–Kutta or Nordsieck methods, can be used. Although these methods are more stable, they can be time-consuming to implement. A fourth-order Runge–Kutta method would require four force evaluations per timestep (not three as in the case of Hamiltonian systems). The book by Lambert (1991) contains a description of many algorithms.

8.3 Neighbour lists for short-ranged potentials

If the interaction forces between the particles are derived from potentials that are short-ranged it is clearly inefficient to examine all pairwise interactions. The computing time is proportional to N^2 for pairwise interaction potentials and to N^3 for three-body potentials or many-body bond-order potentials of the type discussed in Chapter 3. Instead it is more computationally efficient to use a neighbour list, which is periodically updated. Between updates of the neighbour list the program does not calculate the contribution to the force on atom i from the remaining $(N-1)$ atoms but only that from those appearing on the neighbour list of atom i. The use of these neighbour lists therefore considerably reduces the amount of computing time required to calculate the forces on the atoms.

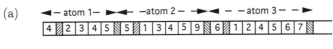

Fig. 8.2 (a) The one-dimensional array NEBLIST. (b) The modification to NEBLIST using the triangle logic, applicable to pair potentials. Only atoms with numbers greater than atom i appear on the neighbour j list of atom i.

The basic idea of the neighbour list is depicted in Figure 8.1. The interatomic potential function for atom i is assumed to cut off at some distance r_c. The first step in the simulation is to construct a list of neighbours for atom i, which lies within a slightly larger sphere of radius $r_l > r_c$. The program cycles over all atoms on the neighbour list i.e. those within a distance r_l of i but only calculates the force on atom i from those particles within a distance r_c of i. This outer shell is important and is arranged to be of sufficient thickness that a particle such as 12 in Figure 8.1, which lies outside the r_l sphere, cannot travel a distance that would allow it to penetrate inside the r_c sphere. Thus, the list should be updated whenever the sum of the magnitudes of the two largest displacements of atoms D_m since the last list update exceeds $r_l - r_c$. Using neighbour lists has implications for the choice of numerical integration timestep and hence the algorithm that should be used in the calculations. High-order algorithms would normally allow larger timesteps to be used for a given accuracy. However, if the timestep is also controlled by the condition $D_m < r_l - r_c$, then the extra work of a higher-order algorithm may not be required. In practise, there is often little to be gained in terms of accuracy by using a high-order method and low-order integration algorithms are generally preferred.

Figure 8.2(a) illustrates a typical neighbour list represented by a one-dimensional array NEBLIST. For each atom i, a pointer KNEB(i) points to the position in NEBLIST which represents the number of neighbours of atom i. Thus, in the above example, KNEB(2) = 6 and NEBLIST (KNEB(2)) = 5 is the number of neighbours of atom 2. The program then calculates the interaction forces and potentials between atom 2 and its five neighbours 1, 3, 4, 5 and 9, cycling between NEBLIST (KNEB(2) + 1) and NEBLIST (KNEB(2) + 1 + NEBLIST (KNEB(2))). Of course it skips the calculation of the potential and forces if any of the neighbours are outside the r_c sphere.

For pair potentials, saving can be introduced by using a triangular summation. Since the force $\boldsymbol{F}_{ij}$ exerted on atom i by atom j is equal to $-\boldsymbol{F}_{ji}$, we need only cycle over atoms j where $j > i$. The modification to the neighbour list in this case is shown in Figure 8.2(b). This speeding up cannot be invoked for many-body potentials.

8.4 Construction of the neighbour lists

For a small number of atoms in a simulation, the neighbour lists can be updated simply by checking the distances between atom i and the remaining $N-1$ particles. Again, a triangular looping technique can be used with pair potentials for speed.

The neighbour list is usually constructed by a two-stage process. First a square neighbourhood test is executed such that limits in the Cartesian directions are defined, for example, $XMIN = X(i) - r_l$ $XMAX = X(i) + r_l$ in the x direction, then, for each i atom, located at $(X(i), Y(i), Z(i))$, the remaining j atoms are tested and kept as possible neighbours if $XMIN < X(j) < XMAX$. A similar process is carried out for the other two Cartesian directions.

For those j atoms which pass the square neighbourhood test, a second spherical neighbourhood list can be used to reject all j atoms for which

$$(X(i) - X(j))^2 + (Y(i) - Y(j))^2 + (Z(i) - Z(j))^2 > r_l^2.$$

Updating the neighbour lists is very time-consuming and can be considerably speeded up if the atoms are sorted in increasing values of their Cartesian co-ordinates. This is known as the 'method of lights' algorithm and can be applied to all co-ordinate directions. A system of N particles can be arranged in increasing values of their x co-ordinates and an array $LOCX$ defined so that $LOCX(i)$ is a pointer to the position in the ordered list of atom i. For example, four particles $\{1, 2, 3, 4\}$ whose x co-ordinates are $\{1.4, 0.7, 2.8, 0.0\}$ would have corresponding $LOCX$ values given by $LOCX(1) = 4, LOCX(2) = 2, LOCX(3) = 1$ and $LOCX(4) = 3$, since particle 4 has the smallest x co-ordinate, particle 2 the second smallest, etc. For a given position j in the sorted list $LOCX(j)$ is the atom number corresponding to j. The neighbourhood test in the x direction then proceeds by increasing j in steps of 1 to a value $j + JT(j)$ when $X(LOCX(j + JT)) - X(LOCX(j)) > r_l$. Similarly, the index j is decreased to $j - JB(j)$ until $X(LOCX(j)) - X(LOCX(j - JB)) > r_l$. Only those atoms in the list lying between $j - JB$ and $j + JT$ are considered as candidates for neighbours of atom $LOCX(j)$. A similar sorting can be done for the y and z co-ordinates, in parallel with the x sort if using a parallel processing machine. The final spherical neighbourhood test could be carried out as

(a)

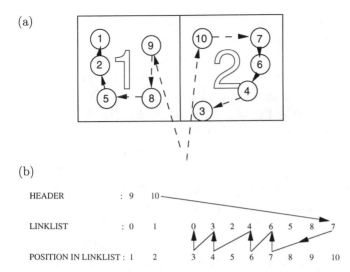

(b)

HEADER   : 9   10

LINKLIST   : 0   1   0   3   2   4   6   5   8   7

POSITION IN LINKLIST : 1   2   3   4   5   6   7   8   9   10

Fig. 8.3 A two-dimensional cell structure. (a) The arrangement of particles in cells 1 and 2 showing the linked list structure through the cells. (b) The linked list structure showing the path through cell 2 of (a).

before. In practise, however, it is easier to use rectangular neighbourhoods but ignore the interactions from particles at distances of greater than r_c when calculating the forces. Although the method of lights is faster than the Verlet list method, it is still not the most efficient method on serial machines. The linked-list method described in the next section is generally to be preferred.

8.5 The cell index method

An alternative method of calculating neighbours for large systems is the cell index method. Here, the simulation region is divided into a set of cubic sub-regions. These cubic cells are chosen so that the side of each cell is greater than r_l. The particles in the system are sorted into each cell, which can be done rapidly at each timestep and implemented using the method of linked lists.

The linked-list method involves sorting all the particles into their appropriate cells. During the sorting, two arrays are created. These are the first element in the list for each cell, the header and the list emanating from the header. The method is best illustrated by an example. Such an example is shown in Figure 8.3(b) for the particles arranged in cells 1 and 2 of Figure 8.3(a). The header for cell 2 is particle 10. The path is determined by observing that the particle 7 is at the tenth position, particle 6 at the seventh position, particle 4 at the sixth, particle 3 at the fourth and a zero at the third position. The zero indicates that there are no more particles and so we must move to the header of the next cell. In three dimensions the computational box (of sides L_x, L_y and L_z) is divided into an $M_x \times M_y \times M_z$ grid of rectangular cells whose sides exceed r_l. An atom can then only interact with atoms in its cell or in the surrounding layer of cells. Each cell then has its sides of lengths L_x/M_x, L_y/M_y and L_z/M_z. It is optimal to have as few atoms in each cell as possible, hence M_x is taken as the largest integer $\leq L_x/r_l$. A similar choice is made for M_y and M_z.

The cell list algorithm lends itself most readily to the application of periodic boundary conditions. It can be extended to free boundaries by making all the cells, on the surface of the computational box, infinite in directions away from the box.

A piece of FORTRAN code that sets up a linked list structure for a system of $M \times M \times M$ cells is given in Allen and Tildersley (1987).

After the linked list structure has been set up and the adjacent rectangular neighbour established, it is still useful to refine the neighbour lists further to spherical regions. The reason for this is that such a spherical neighbour list has volume $\frac{4}{3}\pi r_l^3$. Each rectangular cell has volume $\geq r_l^3$ and there are 27 neighbour cells. The total volume is therefore reduced by a factor of $81/(4\pi) = 6.4$. Thus, using the rectangular cell list requires more than six times as many atoms. Although the triangle logic cannot be used for evaluation of many-body potentials it can be used for setting up the neighbour lists, which means that we only have to search through 14 rectangular cells to obtain the neighbours of an atom, not 28.

The three neighbour list algorithms Verlet, method of lights and linked lists are compared in Figure 8.4. The calculations refer to a C_{60} molecule impacting on a graphite surface, see Figure 8.17. Except for very-small-scale systems, the linked-list algorithm is much faster. For large systems, the theoretical speeds of the algorithms on serial machines are proportional to N^2, $N^{5/3}$ and N respectively, where N is the number of particles in the simulation.

8.6 Timestep control

In atomic collision studies in which the initial kinetic energy is confined to a small number of particles in the system and energy is transferred by collisional processes, it is clearly inappropriate to use a fixed timestep for integrating the equations of motion. Initially, a small timestep is required, which can be increased as the energy in the system disperses. The distance a particle travels during the timestep must not be greater than $r_l - r_c$

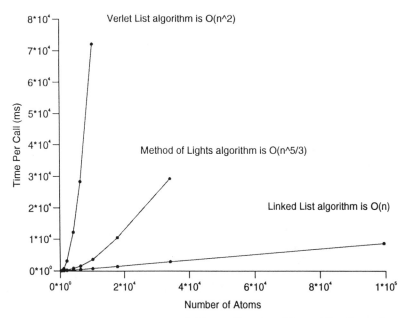

Fig. 8.4 Time comparisons among neighbour list algorithms. The lines through the points are drawn by fitting curves $k_1 N^2$, $k_2 N^{5/3}$ and $k_3 N$ to the last data point.

for the case of spherical neighbour lists. A simple way to control the timestep is to calculate the maximum particle velocity VMAX at each timestep and use it with a distance controller DISMAX, to set the timestep Δt =DISMAX/VMAX. The value of DISMAX is a constant and is best determined by trial and error and set at the start of each computational run. The timestep chosen in this way does not always integrate accurately if DISMAX is too large. An example for which it might fail would be a head-on collision between two particles. Here a small timestep is required when the particles interact with a high potential energy (but a low kinetic energy).

Thus another possibility would be to choose Δt by

$$\Delta t \propto \left(\frac{m}{2\text{EMAX}} \right)^{\frac{1}{2}} \text{DISMAX}, \qquad (8.20)$$

where m is the mass of the particle which has the largest value of EMAX = |potential energy| + |kinetic energy|. At each step a check is made on the maximum distance travelled by each particle. If this exceeds $r_l - r_c$, then Δt should be reduced (usually halved) and the timestep repeated.

8.7 The moving atom approximation

Considerable saving of computer time can be achieved by using a moving atom approximation in atomic collision studies. This is especially useful if we wish to generate statistics from the computer simulation such as the implanted particle distribution or the average sputtering yield but less useful if we are studying the fundamental nature of an individual cascade. In this approximation a list of moving atoms is set up. For a single energetic particle (ion) in collision with a crystal, only the ion is initially on the moving atom list. Other particles are added to the list when the force exerted on them is greater than a certain threshold FRFC. Only the trajectories of these particles which appear on the moving atom list are integrated. The value of FRFC is usually determined by trial and error. Using the neighbour list method of Section 8.3, a suitable algorithm for adding atoms to the moving atom list might be as follows.

First set ITURN = 0, where ITURN is the number of new atoms set into motion by atom i. Cycle through each of the neighbours j of atom i. Check whether FRFC exceeds the specified threshold. If it does, then set

    ITURN=ITURN+1
    NTURN(ITURN)=j

Let NNOW be the number of moving atoms and KSV the length of the array NEBLIST. The moving atoms are stored in the array MVEC, where MVEC(k) is the actual atom number of the kth particle set into motion. A suitable loop to add the new particles set into motion by atom i to the list of moving atoms might be as follows:

    DO 100   INEW=1, ITURN
    j=NTURN(INEW)
    NNOW=NNOW+1
    MVEC(NNOW)=j
    KSV=KSV+1
    KNEB(NNOW)=KSV
    CALL NEWNEB(j)
100 ENDDO

In the above loop, KSV is the next position in the neighbour list when the number of neighbours of atom j is located. The actual number of neighbours and their locations are calculated in the NEWNEB routine. KNEB has the same meaning as in Section 8.3.

The moving atom approximation can be especially useful in sputtering studies in which large statistics are needed and Table 8.2 shows the relative times for sputtering simulations from Ar ions incident normally on a Si{110}

Table 8.2. *Relative computing times with FRFC* $= 1.12 \times 10^4 N$ *(Smith, Harrison and Garrison, 1989)*

Energy (eV)	No of air impacts	CPU time per trajectory
1000	300[a]	100%
1000	5400	30%
400	300	26%
200	300	20%
100	300	11%

[a] FRFC $= 0$

crystal face. The crystal contains 1470 atoms arranged in seven layers. In these simulations the trajectories were run until no more particles were ejected from the crystal surface. Using the moving atom approximation, at 1 keV the sputtering calculations are more than a factor of three faster. The 100 eV trajectories are nearly ten times faster than those at 1 keV when all particles are considered to be in motion.

8.8 Boundary conditions

For high-energy atomic collision events such as kilo-electronvolt particle bombardment of crystals, free boundaries are often preferred to periodic boundary conditions. This is because, if the crystal size for the simulation is chosen to be too small, then it is preferable that energetic particles should leave the sides and bottom of the crystal, rather than reappear from the side opposite from which they exit. For simulations of deposition, on the other hand, periodic boundary conditions across the planes perpendicular to the surface are to be preferred because the edge effects can be important. The Verlet neighbour lists without the linked-list structure can be programmed more easily with free boundary conditions. The linked-list approach adapts most easily to periodic boundary conditions and in the case of free boundaries needs to be modified slightly to include semi-infinite rectangular cells at each crystal edge.

For simulations of deposition, during which energetic particles constantly arrive at the surface, the energy of the system, in the absence of damping forces, must increase. This is usually undesirable and so a model is required that will dissipate excess energy in a realistic way. This is usually done through the boundaries of the crystal and, if periodic boundary conditions have been imposed across the sides of the crystal, it is natural to use the bottom surfaces as the means to this end. For example, Srivastava, Garrison and Brenner (1989) in simulations of silicon MBE used a rectangular crystal consisting of ten layers, each of 32 atoms. In these simulations the atoms remain trapped in their binding potential throughout the simulation. In the first five layers the atoms are allowed to move according to forces derived directly from the interaction potential. The bottom layer is held fixed and the four layers between are used as the heat sink. This is done by introducing a friction term into the equations of motion to maintain the thermal equilibrium of the entire system. The deposition rate has to be adjusted so that the friction term allows the sample to reach the desired temperature before the next atom is deposited, otherwise melting occurs.

There are various constraint methods available for maintaining a set of particles at constant temperature or pressure and a good review is given in Allen and Tildersley (1987). The simplest method is to scale the velocities at each time step by a factor of

$$(T_0/T)^{1/2}, \tag{8.21}$$

where T is the current kinetic temperature at T_0, the desired temperature. This is fairly crude and a more subtle approach would be to use a factor (Berendson *et al.*, 1984)

$$\left[1 + \frac{\Delta t}{t_T}\left(\frac{T_0}{T} - 1\right)\right]^{1/2}. \tag{8.22}$$

Here Δt is the integration timestep and t_T a pre-set time constant. This method forces the system towards the desired temperature at a rate t_T while perturbing the forces much less than does (8.20).

However, a smoother method still is to solve a modified set of equations in the constrained region:

$$m^i\frac{\mathrm{d}^2\boldsymbol{r}^i}{\mathrm{d}t^2} = \boldsymbol{F}^i - \xi m^i\frac{\mathrm{d}\boldsymbol{r}^i}{\mathrm{d}t}, \tag{8.23}$$

where $\boldsymbol{F}^i$ is now the force on the ith particle derived from the interaction potential and

$$\xi = \begin{cases} \dfrac{1}{2t_T}\left(1 - \dfrac{T_0}{T}\right) & T > T_0 \\[2mm] 0 & T \le T_0 \end{cases} \tag{8.24}$$

is a friction coefficient that reduces the temperature of the system at a rate dependent on t_T until $T = T_0$.

8.8.1 Constant temperature–constant pressure molecular dynamics

Evans and Morris (1983) gave the modified equations for the isothermal and isothermal–isobaric ensembles. In the case of N identical particles of mass m the temperature T of the system is given by

$$\frac{3}{2}NkT = \frac{1}{2m}\sum_{i=1}^{N}\left(\boldsymbol{p}^i - \boldsymbol{p}^0\right)^2, \qquad (8.25)$$

where $\boldsymbol{p}^i = m\,\mathrm{d}\boldsymbol{r}^i/\mathrm{d}t$ and $\boldsymbol{p}^0 = (1/N)\sum_{i=1}^{N}\boldsymbol{p}^i$ and, by applying the isothermal constraint using the Gauss principle, the isothermal equations of motion become

$$\frac{\mathrm{d}^2\boldsymbol{r}_i}{\mathrm{d}t^2} = \boldsymbol{F}^i - \alpha\left(\boldsymbol{p}^i - \boldsymbol{p}^0\right), \qquad (8.26)$$

where

$$\alpha = \sum_{i=1}^{N}\boldsymbol{F}^i\cdot\left(\boldsymbol{p}^i - \boldsymbol{p}^0\right) \Big/ \left(\sum\left(\boldsymbol{p}^i - \boldsymbol{p}^0\right)^2\right) \qquad (8.27)$$

and $\boldsymbol{F}^i$ is the force on the ith particle derived from the interaction potential. The total linear momentum is a constant of the motion, as is the temperature T.

The isothermal–isobaric equations of motion are shown to be

$$\frac{\mathrm{d}\boldsymbol{r}^i}{\mathrm{d}t} = \frac{\boldsymbol{p}^i}{m} + \varepsilon\boldsymbol{r}^i; \qquad (8.28)$$

$$\frac{\mathrm{d}\boldsymbol{p}^i}{\mathrm{d}t} = \boldsymbol{F}^i - \varepsilon\left(1 + \sum_{i=1}^{N}\boldsymbol{F}^i\cdot\boldsymbol{p}^i \Big/ \sum_{i=1}^{N}\left(\boldsymbol{p}^i\right)^2\right)\boldsymbol{p}^i. \qquad (8.29)$$

Here, ε is the dilation rate, related to changes in the volume V occupied by the particles:

$$\varepsilon = \frac{1}{3V}\frac{\mathrm{d}V}{\mathrm{d}t}. \qquad (8.30)$$

Equations (8.25)–(8.27) for isothermal MD and equations (8.28)–(8.30) for isothermal–isobaric MD cannot be derived from a Hamiltonian. These equations must be numerically integrated using the algorithms described earlier in this chapter for non-Hamiltonian systems. Isothermal–isobaric ensembles arise mainly in the study of liquids and gases but these constraints could also be applied in crystal deposition simulations.

8.9 Electronic energy losses in MD

Electronic energy loss models are described in detail in Chapter 4. All can be adapted for use in MD simulations. Here we describe how two methods, those of Firsov (1959) and of Caro and Victoria (1989), can be adapted for use in MD simulations. In the Firsov model, a technique due to Wedepohl (Beeler, 1985 p. 105) can be used to obtain the force due to electronic energy loss between two atoms A and B as

$$F_B(=-F_A) = (8.06172 \times 10^{-2})Z^2 \exp{(-C)} \sum_{m=0}^{7} \frac{C^m}{m!}. \tag{8.31}$$

in eV Å^{-1}, where Z is the atomic number of the two particles and

$$C = 13.4646Z^{1/2}|r_A - r_B|^{1/4}. \tag{8.32}$$

In this model, the force is assumed pairwise additive between collision partners and this term is added to the force term due to the interatomic potential.

In the model of Caro and Victoria, the Lagrangian equations of motion (Heermann, 1990) are solved in the form

$$m^i \frac{d^2 r^i}{dt^2} = F^i + \eta(t) - \beta \frac{dr^i}{dt}. \tag{8.33}$$

Here, F^i is the force derived from the interaction potential, η is a random force and β a constant measuring the strength of coupling to a thermal bath of temperature T_0.

The random force η is assumed to be determined from a Gaussian probability distribution function

$$P(\eta) = (2\pi\bar{\eta}^2)^{-1/2} \exp{[-\eta^2/(2\bar{\eta}^2)]}. \tag{8.34}$$

The average value $\bar{\eta}^2 = BkT_0/\tau$, where k is Boltzmann's constant and τ is the relaxation time for electron–phonon interactions, typically $\tau \simeq 10^{-11}$ s. Caro and Victoria give an empirical formula for B in the form

$$B = A \log_{10}{(a\rho^{1/3} + b)}, \tag{8.35}$$

where ρ is the electron density, $a = (3\pi^2)/[3\hbar^2/(e^2 m)] = 3.09a_B$ (a_B is the Bohr radius), A is a fitting parameter close to the value $2Z^2 e^4 m^2/(3\pi\hbar^3)$ and b another fitting parameter taken as 0.65 for Cu–Cu collisions. This form of electronic energy loss is useful when the interatomic potential function is given in terms of ρ, such as in the Finnis–Sinclair or EAM potentials defined in Chapter 3.

8.10 Lattice generation

Lattice generation is straightforward for infinite lattices or semi-infinite crystals that are truncated parallel to a cell edge. All that is required in these cases is to arrange the atoms appropriately in the basic crystal cell and repeat this structure as many times as is required for a realistic simulation, see for example Beeler (1985).

The situation is not quite so straightforward if the crystal surface is not parallel to the basic cell edge and in this section we describe a method for generating a rectangular crystal with a surface plane of arbitrary surface orientation from a cubic lattice. Let the surface plane have Miller indices (l, m, n) with respect to the original cubic lattice, where it is assumed $m \neq 0$. A new co-ordinate system (x', y', z') is defined with the y' direction being perpendicular to the surface plane. For most applications in atomic collision studies the choice of the x' and z' directions is not especially important, so there is a rotational degree of freedom available. In order to simplify the method we choose the following scheme for constructing these directions.

Let the direction cosines of the required surface plane be $(l_2, m_2, n_2) = (l, m, n)/(l^2 + m^2 + n^2)^{1/2}$. If $l_2 = 0$ we choose

$$
\begin{aligned}
l_1 = 0, \quad & m_1 = -n_2, \quad n_1 = m_2, \\
l_3 = -1, \quad & m_3 = 0, \quad n_3 = 0.
\end{aligned}
\tag{8.31}
$$

If $l_2 \neq 0$ then we choose

$$
\begin{aligned}
l_1 = \frac{-nl}{(l^2 + m^2)^{1/2}}, \quad & m_1 = \frac{-mn}{(l^2 + m^2)^{1/2}}, \quad n_1 = (l^2 + m^2)^{1/2}, \\
l_3 = \frac{-m}{(l^2 + m^2)^{1/2}}, \quad & m_3 = \frac{l}{(l^2 + m^2)^{1/2}}, \quad n_3 = 0.
\end{aligned}
\tag{8.32}
$$

The direction cosines $(l_1, m_1, n_1), (l_2, m_2, n_2)$ and (l_3, m_3, n_3) define mutually perpendicular directions. The new co-ordinates (x', y', z') are related to (x, y, z), the original Cartesian directions parallel to the cubic lattice sides, by

$$
\begin{aligned}
x' &= l_1 x + m_1 y + n_1 z, \\
y' &= l_2 x + m_2 y + n_2 z, \\
z' &= l_3 x + m_3 y.
\end{aligned}
$$

The planes parallel to the surface are a distance d_y apart, where $d_y = (l^2 + m^2 + n^2)^{-1/2}$. For a crystal of depth y_d, there are therefore $y_d/d_y = N_Y$ planes. For a fixed depth y' below the surface, the above equations can be manipulated to show that

$$x' = (-ny' + z)\left(l^2 + m^2\right)^{-\frac{1}{2}},$$

$$z' = \frac{1}{m}\left[(ly' - x)(l^2 + m^2)^{1/2} - lnx'\right].$$

(8.33)

Thus, for a rectangular lattice whose sides are

$$0 \le x' \le x_d, \quad 0 \le y' \le y_d, \quad 0 \le z' \le z_d,$$

(8.34)

z ranges from ny' to $ny' + (l^2 + m^2)^{1/2}x_d$ and x ranges from $ly' + (l^2 + m^2)^{-1/2}(-lnx' - mz_d)$ to $ly' - (l^2 + m^2)^{-1/2}lnx'$ for each x'. The quantity y' is incremented in steps of the plane spacing, d_y.

8.11 Lattice vibrations

The models described in this section can also be used in conjunction with binary collision codes. The thermal displacements of the target atoms in the direction s relative to their equilibrium positions in a static lattice can be approximated by the probability density distribution of a quantum mechanical harmonic oscillator. This is a Gaussian distribution of the form

$$P(s) = (2\pi\bar{u}_s^2)^{-1/2}\exp\left[-s^2/(2\bar{u}_s^2)\right].$$

(8.35)

The mean-square vibrational displacements $\bar{u}_s^2$ in the direction of s can be determined either experimentally or by the expression (Radi, 1970)

$$\bar{u}_s^2 = \frac{\hbar^2}{km\theta_E}\left(\frac{1}{\exp\left(\theta_E/T\right) - 1} + \frac{1}{2}\right),$$

(8.36)

where T is the temperature of the target in kelvins, $\theta_E = \theta_D/\sqrt{3}$, θ_D is the Debye temperature, k is Boltzmann's constant and m is the mass of the vibrating atom. This formula is a fit to measured data but does not cover the whole temperature range 20–300 K with a single Debye temperature. The high-temperature form of this expression is

$$\bar{u}_s^2 = \frac{3\hbar^2 T}{km\theta_D^2},$$

(8.37)

which is in agreement with the mean square displacements of Debye–Waller theory

$$\bar{u}_s^2 = \frac{3\bar{h}^2}{mk\theta_D}\left(\frac{T^2}{\theta_D^2}\int_0^{\theta/T}\frac{x\,dx}{e^x - 1} + \frac{1}{4}\right)$$

(8.38)

when θ_D/T is large.

Table 8.3. *Calculated thermal vibration temperatures parallel* ($\theta_\parallel$) *and perpendicular* ($\theta_\perp$) *to the surface for various metals, calculated from the Morse interaction potential, after Jackson (1974). The superscripts 1 and 2 on $\theta_\parallel$ for the {110} face refer to the components in the* $\langle 001 \rangle$ *and* $\langle 1\bar{1}0 \rangle$ *directions respectively*

Surface	{100}		{110}			{111}	
	$\theta_\perp$	$\theta_\parallel$	$\theta_\perp$	$\theta_\parallel^1$	$\theta_\parallel^2$	$\theta_\perp$	$\theta_\parallel$
Cu	192	292	191	303	181	196	328
Ag	142	203	142	209	137	147	226
Au	110	152	111	156	106	116	168
Ni	225	347	225	360	212	230	389
V	235	265	227	420	363	229	241
Mo	239	247	250	402	340	248	255
W	181	191	186	308	263	184	191

For low temperatures the above expression has the asymptotic form

$$\bar{u}_s^2 = \frac{3\hbar^2}{mk\theta_D}\left(\frac{\pi^2 T^2}{60\theta_D^2} + \frac{1}{4}\right). \qquad (8.39)$$

At a surface, the vibrational amplitudes are significantly greater than those of the atoms in the bulk of the solid due to the asymmetric forces experienced by the atoms at the surface. Jackson (1974) evaluated the surface Debye temperature by using the formula for an oscillator

$$\theta_D^s = \left(\frac{\hbar}{k}\right)\left(\frac{q_s}{m}\right)^{1/2}, \qquad (8.40)$$

where q_s is the derivative of the force in the s direction at the equilibrium position of the atom. The Morse potential is used to calculate q_s. Some calculated values for these quantities are given in Table 8.3. It is expected that the values given in Table 8.3 could be improved by using the many-body potentials for metals described in Chapter 3.

In atomic collision studies the thermal vibrational velocity is often not required since this is generally very much less than the speed of any incoming particle. Thus, the undisturbed positions of the crystal can be assumed to be frozen during an energetic collision cascade.

Once the probability distribution for the displaced atoms has been set, it is still necessary to choose the displacements from this distribution. One

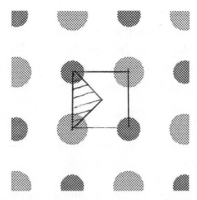

Fig. 8.5 The irreducible symmetry zone for the fcc {010} face, at normal incidence (triangle) and at general incidence (square).

such way to do this is via the Box–Muller algorithm (Press *et al.*, 1992). This transforms random deviates on the interval $(0, 1)$ to values distributed according to the normal distribution.

8.12 Ensembles of trajectories

In atomic collision studies of situations in which energetic particles are incident on the surfaces of materials, experimental data are obtained as a result of large numbers of individual collision cascades. Thus, an individual trajectory cannot be directly compared with experimental data but a properly constructed set of trajectories can serve as a model of an experimental system. If the simulation involves a crystalline target, the atoms in the surface reflect the underlying symmetry of the crystal. This symmetry may be blurred by thermal displacements of the atoms from their lattice sites.

Because a crystalline target surface has symmetry, any point on the surface plane can be mapped onto a minimum set of points, which is called the representative area or irreducible symmetry zone (ISZ). The ISZ depends upon both the crystal and the direction of the incoming particle. Figure 8.5 shows the ISZ for particles across the fcc {010} face. For simulations involving ion scattering or sputtering studies, large numbers of trajectories may have to be run on the computer before good statistics can be obtained. The determination of some physical quantities requires more sampling than does the determination of others. For example, ion back-scattering is a relatively rare event and to obtain good statistics for the angular distributions of the scattered ions may require tens of thousands of trajectories. On the other hand, good statistics for energy distributions of ejected particles requires somewhat fewer.

(a)   (b)

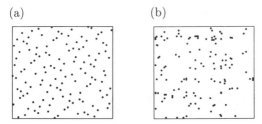

Fig. 8.6 (a) The 128 point Halton sequence mapped on to the unit square. (b) 128 randomly chosen points in the unit square.

It is important therefore to devise a good method for sampling the ISZ. Randomly choosing points over the ISZ is a relatively inefficient way of proceeding. Instead we require a set of points that will be distributed spatially in a more uniform way. The same problem is encountered in Monte Carlo numerical quadrature and we borrow an idea from there. One such way of speeding up the calculation by sampling fewer points is to use quasi-random sequences. One such sequence, the Halton sequence (Halton, 1960), seems especially appropriate. This sequence is defined by

$$u_{i+1} = (\phi_{p_1}(i), \phi_{p_2}(i)),$$

where p_j are co-wise coprime, e.g. $p_1 = 2$ and $p_2 = 3$, and $\phi_p(i)$ is the radical inverse function of i obtained by writing i to base p and 'reflecting about the decimal point'. Thus, for example 15 (base 10) = 120 (base 3) so $\phi_3(15) = 0.021 = \frac{2}{9} + 1/27 = 8/27$. Figure 8.6(a) shows the first 128 points of the two dimensional Halton sequence mapped onto the points (α, β) over the unit square $0 \le \alpha \le 1$, $0 \le \beta \le 1$ with $p_1 = 2$ and $p_2 = 3$. This can be compared with a 128 point sequence whose co-ordinates are chosen randomly to lie between 0 and 1. It can be seen that those points chosen from the Halton sequence cover the unit square 'more uniformly' than do those chosen randomly in Figure 8.6(b).

In ion scattering calculations, the sampling can be even more involved since large portions of the ISZ contribute nothing to the statistics. The trick here is to sample preferentially those areas that do contribute.

If the points are to be distributed over a parallelogram whose vertices have co-ordinates $(x_1, z_1), (x_2, z_2), (x_3, z_3)$ and (x_4, z_4) then the points are distributed according to the formula

$$x = \alpha(x_2 - x_1) + \beta(x_3 - x_1),$$
$$z = \alpha(z_2 - z_1) + \beta(z_3 - z_1),$$

but if it is the triangle with vertices $(x_1, z_1), (x_2, z_2)$ and (x_3, z_3), the same formula can be used, either ignoring points that lie outside the triangle or reflecting those that do about (x_4, z_4) according to

$$x = \alpha(x_3 - x_4) + \beta(x_2 - x_4),$$
$$z = \alpha(z_3 - z_4) + \beta(z_2 - z_4).$$

8.13 Applications of molecular dynamics to surface phenomena

The following sections of this chapter describe some of the problems which have been of interest to the authors of this book and their co-workers in the field of surface science and radiation damage. The field of MD simulations is huge and the choice of material here is therefore fairly subjective. MD simulations have been used to study the interactions both of liquids and of particles with liquids and solids. One of the important quantities for liquids that can be predicted with accuracy using MD, even with pair potential interactions, is the structure factor $G(r) = 4\pi^2 \rho(r)$, where $\rho(r)$ is the density of atoms at a distance r away from a given atom. Figure 8.7 shows such a function obtained from MD by heating a Cu crystal of 1445 atoms to 1462 K using a Morse interaction splined to a Born–Mayer potential until equilibration. The comparison with experimental data obtained from neutron diffraction measurements is shown to be in excellent agreement with the MD predictions. This structure factor has also been calculated in the centre of collision cascades and has been used as evidence to suggest that melting can occur during radiation damage (Webb, Harrison and Barfoot, 1985) Many other liquid properties have been simulated and a description of some important applications to liquids is contained in the book by Allen and Tildersley (1987). However, in this chapter we will concentrate on the interaction of particles with solid surfaces.

Simulation of the ejection of atoms from a crystal gives a good example of where computational techniques can be used to predict a large number of experimentally observed quantities but also to probe areas outside experimental observation. The ejection of particles from crystals has been extensively studied using MD, thanks primarily to the pioneering work of Don Harrison. The extra detail inherent in the MD over the BC approach enables the study of such phenomena as pit formation by single-particle impact, cluster ejection or atomic excitation. The use of computer graphics has shown that there are many simultaneous interactions between moving atoms and thus that the BC approximation can be suspect for use in sputtering studies (Webb *et al.*, 1993). In this reference the number of multiple collisions

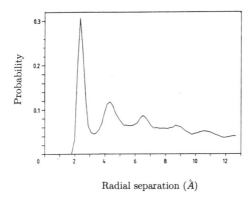

Radial separation (Å)

Fig. 8.7 The radial distribution function calculated using MD after heating a crystal by imparting initial kinetic energy to particles in a Cu crystal lattice.

is calculated for one particular trajectory for 25 keV Ar on Rh{111} and is shown to be a significant proportion of the total number of 'hard' collisions during the early stages of the cascade.

In the following subsections we describe results from MD simulations of sputtering. The results described here have been obtained by using a code based on the QDYN model. The philosophy of this code has been discussed in detail in an article by Don Harrison (Harrison, 1988) which also contains a review of results up to 1988.

8.13.1 Angular distributions of ejected particles

The angular distributions of ejected particles can give important information regarding the structure of the crystal surface from which they were ejected. Crystal structure plays an important role in two ways. First, the yield of ejected particles depends on the beam orientation, so that if the incoming particles are oriented down a channel they deposit most of their energy below the surface region, creating damage within the crystal rather than sputtering the surface. Secondly, a well-defined angular distribution pattern, the so-called 'Wehner spots', arises as a result of the arrangement of surface and subsurface atoms. These patterns can be accurately simulated by MD. For close packed fcc metals, the main concepts that have arisen are that channelling and blocking by surface atoms are responsible for these patterns. Atoms are preferentially ejected in the direction in which the distance between nearest neighbour surface atoms is greatest and suppressed in the direction in which it is least. Figure 8.8(a) shows a calculated spot pattern together with the arrangements of surface atoms for Cu{100}. The ejection patterns, shown in Figure 8.8, are produced by projecting the ejection direction from the origin

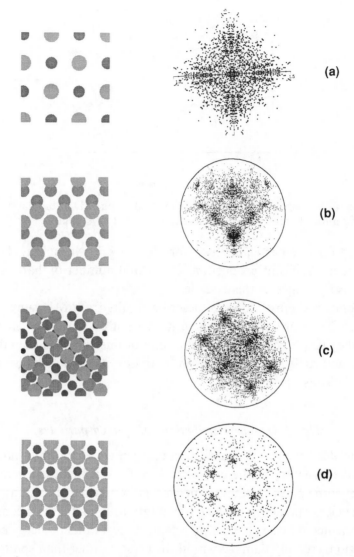

Fig. 8.8 Surface arrangements of atoms together with the ejection patterns calculated by MD: (a) Cu{100} (b) Si{110} (c) Si(2 × 1){100} and (d) graphite {0001}.

onto the unit hemisphere and then projecting the unit hemisphere onto a plane. They are thus not directly equivalent to the experimentally observed phenomenon, which was observed by placing a flat plate above a crystal. For semiconductors, in which the crystal structure is more open, the ejection mechanisms are more complex and the peaks in the angular distributions for the bulk terminated Si{100} face are twisted through 45° relative to the fcc {100} face. The calculated spot patterns together with the crystal structures

Si{110}

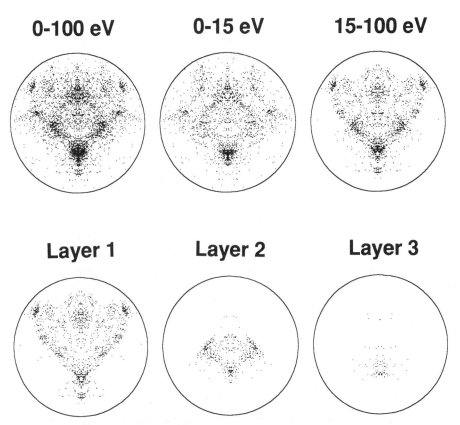

Fig. 8.9 The ejection pattern for Figure 8.8(c) for ejected particles classified according to energy and depth of origin.

for various Si faces are also shown in Figure 8.8. For graphite the hexagonal structure of the lattice is clearly reproduced in the ejection pattern. Figure 8.9 shows a layer by layer analysis of the ejection pattern shown in Figure 8.8 for the Si{100}(2 × 1) face. We see that the open crystal structure allows atoms to escape from the second and third layers. The peak in the second layer yield is along azimuths that are unblocked by first-layer atoms. A full description of these ejection patterns, together with the collision mechanisms that give rise to them, is given in Smith, Harrison and Garrison (1989). The angular distributions of particles ejected from single crystals have been measured and compared with experiment. A direct comparison is shown in Figure 8.10 for the reconstructed GaAs{100}(2 × 4) surface and for Rh{111}.

Table 8.4. *The percentage contribution of the different layers for normal Ar incidence bombardment at 1 keV*

	Si{100}(2 × 1)	Si{110}	Graphite{1000}	Cu{100}
Layer 1	46%	75%	94%	90%
Layer 2	31%	19%	4.5%	9%
Layer 3	16%	6%	1.5%	1%

8.13.2 The depth of origin of ejected particles

The ejection patterns described above also contain depth information. It is possible to analyse these patterns and determine where the ejected particles which make up the pattern originate. This is clearly an important quantity in surface analysis, for which accurate composition–depth profiles are required. Transport theory models predict that ejection can take place from deep within a bulk solid. However, the early MD simulations of sputtering by ion impact at energies up to 1 or 2 keV were in conflict with this hypothesis and experiments later confirmed that the bulk of the ejected material originated from the top atomic layer. Table 8.4 compares the calculated depth distributions of particles ejected from single crystal Cu, Si and graphite, all for normal incidence bombardment at 1 keV.

8.13.3 Ejected atom energy distributions

The energy spectrum dN/dE of ejected atoms also provides a means of direct comparison between the simulation results and experiment. At beam energies above a few hundred electronvolts and for normal incidence upon fcc materials, direct ejection by the ion accounts for only a small fraction of the total ejection yield. The simulations also indicate that dN/dE is sensitive to the direction of atom ejection. Transport theory predicts that the peak in the ejection yield would be at an energy equivalent to half of the cohesive energy of the solid (see Chapter 5). However, simulations carried out by Garrison *et al.* (1988) for Rh{111} show that this is not the case if the embedded atom potential is used to model the atomic binding, although it would be the case were pair potentials to be used. The experiments confirm the higher predicted value of around 5 eV for the peak, see Figure 8.11.

For semiconductors, the peaks in the energy distribution are even higher. Results for the Si{100}(2 × 1) face show a peak of 8 eV, nearly double the cohesive energy of Si. The simulation results for graphite show a flat

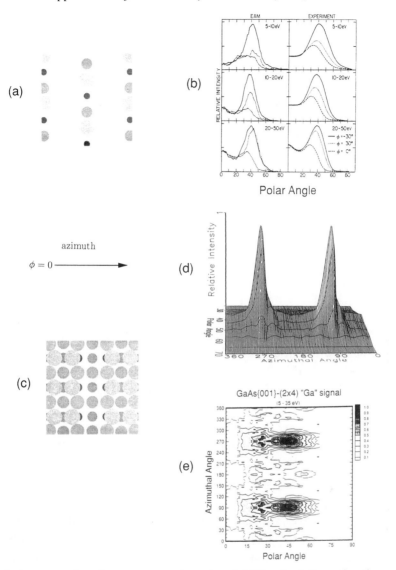

Fig. 8.10 A comparison between experiment and MD calculations for the angular distributions of sputtered particles for a normally incident Ar beam at 1 keV. (a) The arrangements of surface atoms for Rh{111}. (b) A comparison between the experimental and MD simulation results (courtesy B. J. Garrison and N. Winograd). (c) The arrangement of surface atoms for GaAs{100}(2 × 4). (d) The experimental Ga angular distributions. (e) Results of MD calculations.

distribution up to about 20 eV. However, it must be stressed that these results are for single crystals. Ejection from a damaged or amorphised surface is more likely to be in accord with Boltzmann statistics.

The energy distributions also depend on the angle of incidence of the

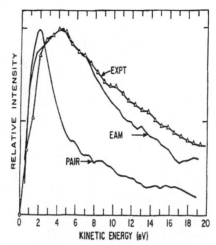

Fig. 8.11 A comparison between experiment and MD calculations for the energy distributions of particles sputtered from Rh{111} (courtesy B. J. Garrison and N. Winograd).

incoming beam. At normal incidence for 1 keV bombardment of Si, no particle was found to be ejected with an energy >100 eV; 88% of ejected particles were emitted with energies less than 50 eV and 59% with energies less than 25 eV. The incidence angle corresponding to the maximum yield, θ_{max}, for Si{110} was calculated to be 72.5° for $\phi = 0$° (Smith, Harrison and Garrison, 1990). At this angle the calculated peak of the ejection energy is greater than it is at normal incidence. Statistics from 300 trajectories at 72.5° incidence show that there is little variation in the distribution of low-energy ejected atoms between about 4 and 20 eV. Analysis of the data shows that 56% of ejected particles are now emitted with energies >25 eV and 22% with energies >100 eV. They are also scattered preferentially in the forward direction. Although the results show that a large number of high-energy atoms are ejected, the increase in the yield at $\theta_{max} = 72.5$° compared with that at normal incidence is also made up from a large increase in the number of low-energy atoms. The yield variation with incidence angle is discussed further in Section 8.13.5.

8.13.4 Atoms per single ion (ASI) distributions

Experiments measure average ejection yields. MD simulations, on the other hand, obtain statistics from a finite set of trajectories incident over an area of the surface that is representative of the crystal as a whole, usually an irreducible symmetry zone. At beam energies of about 1 keV, the atoms per

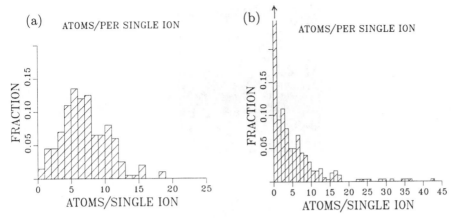

Fig. 8.12 The ASI distributions for (a) 1 keV Ar on Cu{11 3 1} (b) 20 keV Ar on Cu{11 3 1}.

single ion distribution for fcc materials can extend to around five times the average yield. There are also trajectories along which the ion channels and deposits its energy deep within the crystal ejecting no particles. These effects become more pronounced as the beam energy increases. For example, for Cu the average measured sputtering yield for 1 keV Ar bombardment is around four atoms per ion. This rises to seven atoms per ion at 20 keV (Behrisch, 1981). The simulation results using pair potentials for Cu show a single peak distribution at 1 keV for the {11 3 1} face, see Figure 8.12. However, as the energy increases, more ions implant into the crystal causing no sputtering until at 20 keV a significant contribution to the ejection yield comes from relatively rare events that sputter large numbers of particles and cause pits. At this energy the simulations show that yields of over 100 atoms can occur for some impact points. Such events have been termed sputtering 'mega-events'. These mega-events arise when most of the energy of the incoming ion is transferred to near-surface atoms and the resulting momentum has only a small downward component. This example of large-yield events illustrates the inapplicability of Boltzmann statistics to sputtering studies and can give clues about how surface micro-topographical features can develop. Such features severely affect the resolution of composition–depth profiles in surface analysis. Experimental evidence from the scanning tunnelling microscope has confirmed the simulation results that single-particle impacts can form craters (Wilson, Knipping and Tsong, 1988). The overlap between craters formed from successive mega-events could be the first steps towards the formation of larger topographical features.

STM results from low energy (≤1 keV) Ar bombardment of graphite have

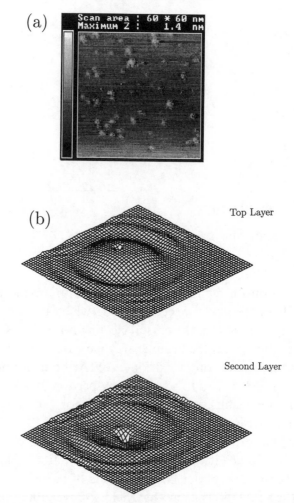

Fig. 8.13 (a). A 60 nm × 60 nm STM image of graphite bombarded by a 500 eV Ar ion at normal incidence. (courtesy of Dr B. V. King). (b) A computer simulation of the development of a bump under the same bombardment conditions, 0.6 ps after impact. The height of the bump is 1 Å.

indicated the presence of bumps on the surface after bombardment, rather than craters. The MD results have shown that these are due to interstitial carbon atoms between the first and second layers, which occur due to the passage of the energetic Ar atom. The results show a similar inward expansion of the second layer. The sizes of these bumps (1–5 Å in height) predicted by the MD simulations are in close agreement with the experimental measurements. Figures 8.13(a) and (b) compare the experimental results with the MD calculation.

8.13.5 Yield variation with incidence angle: impact collision SIMS

Because of the well-defined crystal structure, the sputter yield curves exhibit peaks and troughs as the incidence angle varies. The positions of the peaks and troughs can be predicted by a simple two-atom interaction model in many cases, so that a full MD or BC simulation is unnecessary. The positions of the troughs are just the channelling directions but Chang and Winograd (1989) showed that, if the ion beam is incident in the correct direction, then an atom in the top layer may deflect ions towards an atom in a lower layer, thereby increasing momentum transfer into the surface region and increasing the total ejection yield. The focusing of ion incidence flux onto and away from lower layers can be simply modelled in terms of shadow cones, regions of space behind an atom where no flux is incident see Figure 2.11. This technique forms the basis of a technique called impact collision SIMS (ICSIMS) for determining surface structure and is analogous to impact collision ion scattering spectroscopy (ICISS) except that, instead of detecting the scattered primary ions, secondary ejected particles are detected. The two techniques are shown schematically in Figure 2.10. Figure 2.11 shows that there is an enhanced flux region at the edge of the shadow cone. In Figure 2.10(b) the orientation of the beam is such as to give rise to an increased ejection yield where the edge of the shadow cone lies just below a neighbouring lattice particle. If the shadow cone edge directly intersects a neighbouring atom position we would expect to be able to detect an enhanced yield in the backscattered direction. This is the principle of the ICISS technique for determining surface structure and can be analysed with reasonable accuracy using the BC approximation.

The ICSIMS technique has been used to determine the atomic geometry of a number of different crystal surfaces. The technique has to be fairly carefully applied because the shadow technique will not be applicable for all situations and it is therefore wise to compare the results with full MD simulations. Figure 8.14 gives the calculated yield as a function of incidence angle for the Si$\{110\}$ face for 1 keV Ar bombardment along the $\phi = 0°$ azimuth. There are peaks in the $Y-\theta$ curve at $60°, 65°$ and $72.5°$. At an angle of $72.5°$ the forward end of the shadow cone from a surface atom intersects with another surface atom, see Figure 8.15(a) for the shadow cone based on a two-atom collision model. The mechanisms which give rise to the increase in the calculated yield as θ increases from 67.5 to $72.5°$ at 1 keV have been examined. These show that the major part of this increase is due to the ejection of the surface atom located at the forward end of the shadow cone, giving a direct confirmation of the shadow cone yield-enhancement

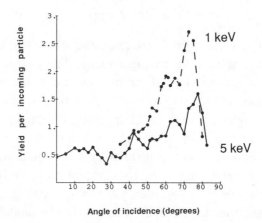

Angle of incidence (degrees)

Fig. 8.14 The ejection yield calculated by MD plotted as a function of polar incidence angle θ for bombardment of the Si$\{110\}$ surface along the $\phi = 0$ azimuth at bombardment energies of 1 and 5 keV.

theory for this peak. Other peaks can be similarly interpreted. The full MD simulations along the symmetry line are shown in Figure 8.15(b). Consider now the $Y–\theta$ curve for the dimer reconstructed surface shown in Figure 8.16. The data for these simulations were obtained by using a target of 1280 atoms. The chosen azimuth of $\phi = 45°$ was because the surface atoms are regularly spaced in this direction and we have shown from the simulations of the $\{110\}$ surface that shadow cone yield enhancement defines a more dominant peak for the surface atoms. Thus, if the model were to be appropriate for the $\{100\}(2 \times 1)$ face, this azimuth ought to be a direction in which we should see peaks due to surface atoms. However, Figure 8.16 shows no hint of structural information of any kind and the yield curve is now representative of the shape characteristic of amorphous materials, rising to a single peak at 65° and then dropping to zero at near grazing incidence (see Figure 9.8). A numerical fit to these data was later used for some topography simulations. (See Chapter 9.) The shapes of the curves are in agreement with experimental data (Zalm, 1983) measured for amorphous Si. The peak at 65° cannot be attributed to shadow cone yield enhancement. The reason for this is that the distance between adjacent dimer atoms is such as to cause the shadow cones to overlap and, in addition, many trajectories pass through the surface layer undisturbed.

8.13.6 Cluster ejection

MD simulations can give quantitative predictions concerning the ejection of clusters, for example, the proportions of dimers and trimers from pure

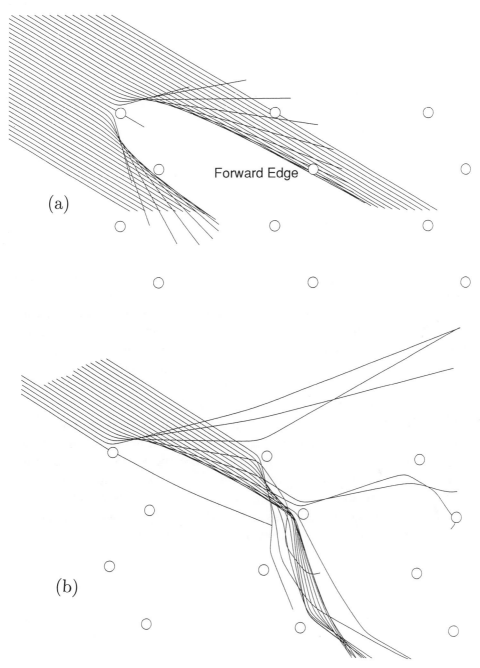

Fig. 8.15 1 keV Ar bombardment of Si{110} at 60° incidence: a comparison of ion trajectories between the single BC event (a) and MD calculations (b) for trajectories initially incident along a symmetry line. The atomic positions in the symmetry plane are shown by the circles. Those in (a) play no part in the calculation and are shown for comparison only.

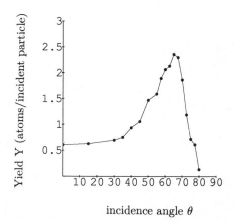

incidence angle θ

Fig. 8.16 The calculated ejection yield as a function of polar incidence angle θ, for the bombardment of the Si$\{100\}(2 \times 1)$ surface.

elemental crystals or of more complex molecules from organic compounds. This is clearly important in SIMS. The quantitative aspects of simulations to date must be treated with caution because the interatomic potentials are not always good enough to describe accurately the correct cluster energetics as well as the bulk material properties. Some of the many-body potentials now being developed do allow for an accurate description both of small cluster and of bulk properties (see Chapter 3), which means that the proportion of particles ejected as clusters should be predictable.

The main focus of simulation in the analysis of clusters has centred around how they are formed i.e. are they ejected from a lattice intact or are they formed by recombination after ejection? Since the SIMS technique relies on the particles being ejected intact, this is a rather crucial question. It appears that both mechanisms are possible for metals, but the dominant mechanism for covalent materials is that the clusters are ejected intact. A video illustrating this process is available from the authors.

8.13.7 Cluster beams

Ion beams such as those that originate from liquid metal ion sources contain a mixture of single atoms and small cluster fragments. Beams that contain extremely large clusters can have startling effects. These clusters can set into motion large numbers of particles within a small volume. Recent work (Smith and Webb, 1993) with the fullerene (C_{60}) molecule has indicated that hexagonal surface waves can be formed on graphite and that the molecule can bounce off surfaces and remain intact at energies of up to

(a)
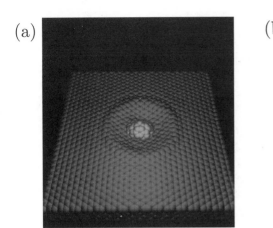

(b)
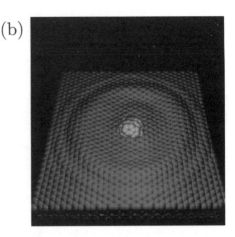

Fig. 8.17 The atomic positions for a 1 keV normally incident C_{60} molecule (a) 262 fs and (b) 532 fs after impact. Note the hexagonal shape of the surface waves produced by the impact.

about 250 eV. Figure 8.17 gives an example of such surface waves. These MD simulations with large molecules have indicated that shock-like disturbances from the impact can transfer small amounts of energy through the lattice. Evidence both from experiment and from simulation shows that rotational and vibrational energy can be transferred to the C_{60} molecule after impact.

The interaction of C_{60} molecules with Si crystal surfaces has also been modelled using MD. In contrast to the results for graphite, it is found that the molecule rarely reflects intact from the surface. When reflection does occurs it is always at near grazing incidence with impact energies less than 300 eV. At normal incidence and similar energies, the molecule remains intact but becomes embedded in the surface layers of the silicon lattice. Grazing incidence (about 75°–80° to the surface normal) at energies of a few hundred electronvolts results in the fullerene molecule becoming trapped in the surface binding potential. The molecule can roll across the surface for up to one revolution before coming to rest. An interaction like this, which occurs between the fullerene molecule and the {110} surface, is shown in Figure 8.18. In this case the molecule sticks and rolls over the surface, pulling silicon atoms from the surface, which stick to the molecule before the molecule itself finally sticks to the surface.

At energies greater than about 500 eV, at grazing incidence, the molecule breaks up on impact, with the majority of the constituent atoms being reflected. Normal incidence, with impact energies in excess of 1 keV, leads to disintegration of the fullerene molecule and sputtering from the crystal, with the ejection of atoms and larger Si_xC_y molecules. This is especially

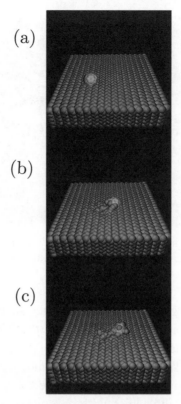

Fig. 8.18 The atomic positions for a 250 eV C_{60} molecule incident at $\theta = 80°$, $\phi = 25°$ on Si{110} after various times: (a) 250 fs, (b) 750 fs and (c) 1250 fs.

evident at energies greater than 4 keV, at which high-energy deposition near the impact point creates a crater surrounded by a 'hot' disordered region from which Si atoms can be thermally ejected for times up to the order of 2 ps. The development of the cross-sectional profile of such a crater is shown in Figure 8.19.

8.13.8 Radiation damage in metals

Radiation damage in metals has been extensively studied using MD by a number of groups. See for example Bacon and Diaz de la Rubia (1994) for a review. Many studies have been carried out by initially imparting a large energy (about 5 keV) to an atom at a central site within a crystal lattice, in a non-channelling direction, and then calculating the resulting cascade. The simulations have shown that point defects produced by the cascades are the result of self-interstitial atoms (SIAs), which are ejected from the core of the cascade by replacement collision sequences. As energy is transferred to the

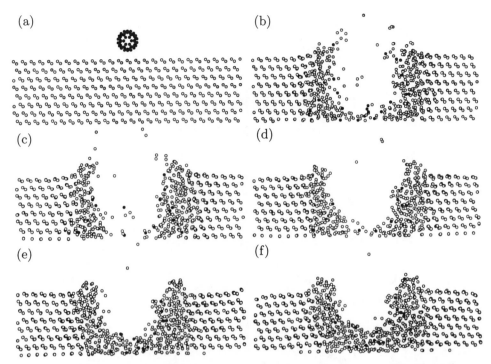

Fig. 8.19 The atomic positions for a 5 keV C_{60} molecule normally incident on Si{111} after various times (a) 0, (b) 400 fs, (c) 800 fs, (d) 1200 fs, (e) 1600 fs and (f) 2000 fs.

stationary atoms all the atoms in the core region can be set in motion so that this region has the characteristics of a molten core. As energy is further dissipated, re-solidification takes place, causing the vacancies to be frozen in the central core region and interstitial clusters outside this region.

If the primary particle is injected into the crystal through a surface in a non-channelling direction, then it is possible for energy to be deposited close to the surface, causing surface damage. Such a situation is described in Figure 8.20. Here a 2 keV Ar ion is injected across the Ni{100} face at normal incidence. After 7.25 ps the damage is annealed out and, surprisingly for this trajectory, there are only adatoms and vacancies left, which form a shallow crater. These calculations were carried out using periodic boundary conditions and no inelastic energy loss. Ackland's Ni potential was used splined to the ZBL potential at close range for a crystal of 30 000 atoms. It is known, however, that the electron–phonon coupling for Ni is high and, if we include this in the model, then the cascade is more damped and the resulting damage is less, with dumb-bell interstitials present away from the impact point.

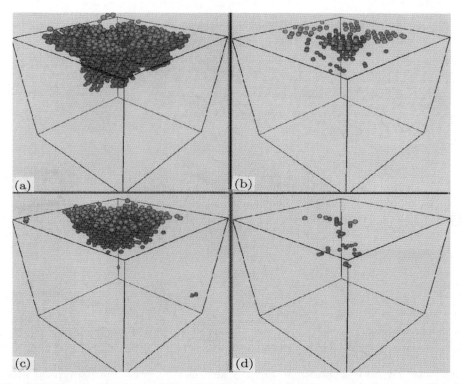

Fig. 8.20 The damage created by a 2 keV Ar atom incident normally on the {100} surface of Ni. Atoms above the surface are shown lighter. The vacancies are the circles, slightly smaller than the interstitals. (a) Showing the peak damage, 250 fs after impact. (b) Partial annealing after 2 ps. (c) The crystal 7.25 ps after impact, showing a crater surrounded by adatoms. (d) A similar trajectory after the damage has annealed out but including inelastic energy losses.

8.13.9 Radiation damage in semiconductors

Figure 8.21 shows the results of such simulations for a 1 keV Si particle injected across the {110} face. The calculations here were carried out with a crystal consisting of 29 108 atoms arranged in 121 layers and the Tersoff Si potential splined to a Molière potential for close particle separation. The damage caused by injection at four separate impact points on the crystal surface is shown. In this direction none of the trajectories give evidence of a cascade consisting of a central vacancy-rich core surrounded by interstitial atoms. Figure 8.21(a) shows a typical trajectory, along which the ion causes some initial damage followed by some channelling before finally causing end-of-range damage and coming to rest. In Figure 8.21(b) intermediate channelling occurs with damage confined to near the channel in which the implanted particle travels. Here it undergoes short-wave oscillations whose

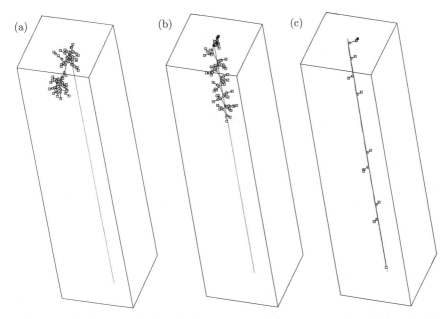

Fig. 8.21 Some examples of collision cascades from calculations carried out for 1 keV Si ions incident normally on Si{110}. The vacancies are plotted as circles and the interstitials as squares. A straight line joins a vacancy created by the displacement of an atom to its final resting place. Only those atoms which are displaced a distance greater than half an interatomic spacing are drawn. The figures correspond to different impact points on the crystal. In (a) we see a trajectory where the ion causes some initial damage before channelling and finally coming to rest. In (b) there is some axial channelling and in (c) deep axial channelling. In all the simulations the same surface area of 68 Å × 68 Å is used.

wavelength decreases before it finally comes to rest. In Figure 8.21(c) the ion passed right out of the bottom of the crystal, causing just single radial displacements every few layers. Finally, Figure 8.21(d) shows a cascade with some evidence of small replacement collision sequences propagating outwards from the path of the implanted particle.

8.13.10 Ion implantation

Recent improvements in computational power mean that some aspects of ion implantation can now be studied using molecular dynamics, at least for low to medium energies. The simulations shown in Figure 8.22 are for 50 eV implantation into Si. This system was chosen because SIMS data had indicated that the ranges of 50 eV B into crystalline Si far exceeded those predicted by transport theory, whereas for amorphous Si they did not. The simulations were carried out assuming a ZBL interaction potential between the B and Si particles. The Si–Si potential for the MD simulation was the many-body Tersoff potential as above. The impacting particle trajectories are shown in Figure 8.22. It can be seen that, for the {110} face, channelling by the B atoms is possible even at this low energy.

8.13.11 *Ion scattering and surface skipping motion*

The scattering of ions from a surface provides examples for which the BC approximation works well and also examples for which it fails. In medium-energy ion scattering (MEIS) the energy and angular distributions of light ions scattered from a crystal surface can be used to determine the structure and composition of the crystal. Results of some MD computer simulations of 2 keV He ion scattering from GaAs{110} illustrate that the trajectories can be approximated by straight lines before and after collision and, because the He particles travel much faster than the crystal atoms that are set into motion, collisions with moving atoms are unimportant. Thus, the BC approximation works well in this case. There are many problems involving light-ion scattering that can be usefully modelled using BC algorithms. The amount of computing time required to generate sufficient data for comparison with ion scattering experiments means that MD simulations are hopelessly inefficient to use for generating energy and angular distribution data.

Where the BC approximation breaks down is at near grazing incidence, especially if there is an attractive part to the interatomic potential. The kinetic energy distributions of atoms scattered at grazing incidence from a clean crystal surface show multiple peaks. Snowdon, O'Connor and Macdonald (1988) interpreted these as due to transient adsorption or 'skipping motion' of the beam, with the highest energy peak representing simple elastic scattering and each subsequent peak an additional vibration in the binding potential. Simulations using pair potentials for Si atoms incident at a polar angle of $\theta = 4°$ from a Cu surface have shown that the combination of thermal vibrations and the chemical binding potential can be responsible for this skipping motion. The crystal size used for these simulations consisted of 7645 atoms arranged in four layers. The crystal volume was a rectangular box whose surface area was assumed to be 325 Å × 32.5 Å. The bombarding Si particles were assumed to be incident at an azimuthal angle $\phi = 0$ to the {110} direction. The incident energy was in all cases 1 keV. The simulations indicate that approximately 5% of Si neutral atoms remain trapped after the first impact and that edge effects can play an important role in the trapping process.

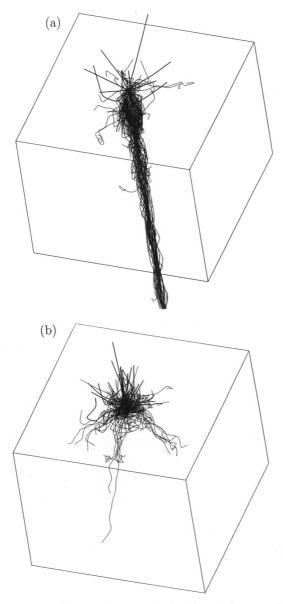

Fig. 8.22 The trajectories for 50 eV normally incident B ions on (a) Si{110} and (b) Si{100}(2 × 1). The depth of the crystal is 13.5 Å and the surface area 27 Å×27 Å.

Figure 8.23 shows some trajectories of a 1 keV Si neutral atom near steps on the Cu{111} surface, at 4° incidence. The simulations show how these steps can cause the Si particle to become trapped in the surface binding potential (Smith, O'Connor and von-Nagy Felsobuki, 1993).

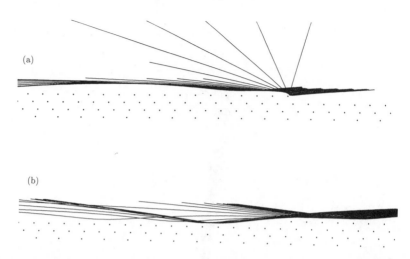

Fig. 8.23 Scattering trajectories near a two-dimensional step on the Cu{111} surface. The initial incidence angle is 4° and the incident Si particle has 1 keV energy: (a) an upward step and (b) a downward step.

8.13.12 Crystal growth

Simulation of growth processes using MD is more difficult to achieve than sputtering because it requires a series of individual impact events to be modelled consecutively. In a real MBE process this occurs quite slowly and so, for the simulation to work in a reasonable timescale, energy has to be artificially extracted from the simulated crystal at a rate much faster than that which occurs in practise. In addition, the size of crystallite that a simulation can study is only small with current computing power. Nonetheless, some interesting experimental phenomena can be modelled and one such success has been the work of Garrison and co-workers on the initial stages of epitaxial growth on the dimer reconstructed Si{100}(2 × 1) face (Srivastava, Garrison and Brenner, 1989). Such a simulation was achieved using a rectangular Si crystal consisting of ten layers of atoms with 32 atoms per layer and periodic boundary conditions parallel to the surface. The bottom layer was anchored and the next four layers forming the thermal bath region using the energy dissipation modelled (see earlier in this chapter). The energy dissipation rate was sufficient to allow equilibration of the system between arriving atoms every 2–3 ps. Deposition of 4–5 monolayers required 800 h of computer time on an IBM 3090. Nonetheless, these simulations have revealed a *novel* mechanism of co-operative motion, whereby dimers simultaneously open in adjacent rows. This mechanism predicts that there should be single rows of dimers rather than isolated dimers in the first epitaxial layer. Concurrently with these calculations, the same isotropic growth of Si{100}

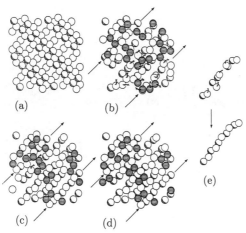

Fig. 8.24 Molecular dynamics simulations of epitaxial growth on Si{100}(2 × 1). (a) The original surface at 800 K. The shaded circles represent the surface dimer atoms. The open circles represent the next two sub-surface layers. Periodic boundary conditions connect the top and bottom of the surface and the left and the right. The dimer rows proceed from the upper left to the bottom right. (b) The surface after deposition of 1.5 ML. The original surface dimer atoms are shown as shaded circles. The sub-surface atoms shown in (a) are no longer displayed. The hatched circles are the topmost deposited atoms and the open circles are the remaining deposited atoms. The curly arrows indicate the directions along which the atoms move in the next 512 ps. The straight arrows on the sides of the crystal are a guide to the periodic boundary conditions and the correlated motion mechanism. (c) The surface after 512 ps of equilibration at 800 K. The atom representations are the same as in (b). (d) The surface after another 476 ps of equilibration. (e) The dimer opening mechanism. The original dimer atoms are the shaded circles and grow into chains as the deposition continues.

during MBE was observed in the scanning tunnelling microscope (Hamers, Köhler and Demuth, 1990). The reaction dynamics indicate that epitaxial growth grows preferentially in a direction perpendicular to the original dimer rows on the surface. As the growth continues the simulations have shown that the top layer constantly reconstructs and the atoms in the next to the top layer constantly return to the bulk sites. Figures 8.24 and 8.25 compare these effects. The success with Si illustrates the power of molecular dynamics simulations as a possible tool for the prediction of the optimum conditions for growth.

The growth of C_{60} on Si{100}(2 × 1) can also be modelled. The simulation is started with a bare Si{100}(2 × 1) surface. Periodic boundary conditions are applied to the sides of the surface, with free boundary conditions in the vertical direction. Two sizes of surface were used. The first surface was approximately 54 Å × 54 Å and the second 46 Å × 46 Å. The latter

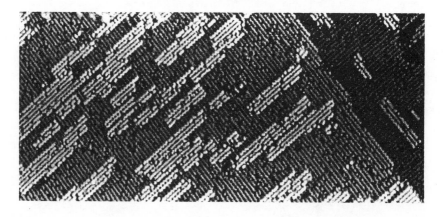

—— 10nm

Fig. 8.25 STM pictures of epitaxial growth on Si{100}(2 × 1).

size was chosen because its periodicity fits exactly a size that would allow either perfect cubic or hexagonal growth to occur. The target is five layers deep but the bottom two layers of the target were kept fixed. Simulations were carried out with the target held fixed both at 300 and at 600 K. Fullerenes were introduced into the simulation, with an internal energy corresponding to the sublimation temperature of 600 K for C_{60}. A further 0.1 eV kinetic energy was added to the centre of mass of each fullerene, with the velocity in a random direction. Experimentally, fullerenes are deposited at a rate of approximately one monolayer per minute. This is far too long a timescale to be simulated by MD. To reduce the simulation time, we introduce a new fullerene either when the molecule has reflected out of the system or 2 ps after binding to the surface. Once fullerenes are on the surface, they are coupled to a thermal bath corresponding to the target temperature. Any reflected fullerenes are removed from the simulation.

Figure 8.26 contains snapshots from a simulation of growth on the 54 Å × 54 Å and the 46 Å × 46 Å surface at 300 K. The fullerenes are mobile in the first layer, especially up and down the dimer rows, and congregate into groups. The combination of the Si–C_{60} interactions and interactions between C_{60} results in the first monolayer on the 54 Å × 54 Å being characterised by mostly cubic stacking but with some hexagonal stacking. This mixture of cubic and hexagonal stacking is in agreement with the STM results. As more molecules accumulate on the surface, they congregate in groups

(a)   (b)

(c)   (d)

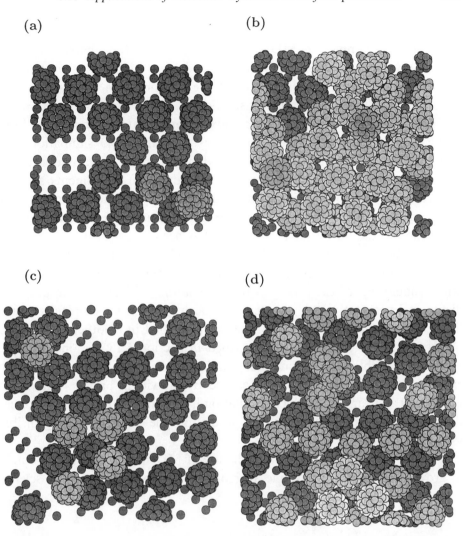

Fig. 8.26 In (a) and (b) various stages in the MD simulation of C_{60} on Si$\{100\}(2 \times 1)$ on a 46 Å $\times$ 46 Å surface at 300 K are shown. Note the mainly close-packed stacking. In (c) and (d) there is growth on a 54 Å $\times$ 54 Å surface at 300 K. Note the mixture of cubic and hexagonal stacking.

and the film becomes more stable. Molecules that impact above a closely packed region in the first monolayer can settle in hollows between molecules. For the 46 Å $\times$ 46 Å surface the stacking is mainly hexagonal. This illustrates the influence that boundary conditions can have in these small crystal simulations.

8.14 Conclusion

Most of the examples of MD simulations given in this chapter have an accompanying video, which can be obtained from the authors on request, but clearly the examples chosen reflect the authors' own interests and by no means constitute a complete list. For example, chemical reactions at surfaces and catalysis have been avoided. Ionic materials have not been studied. Notwithstanding these gaps, it is clear that MD simulations are becoming increasingly attractive because they can provide microscopic detail of chemical and physical phenomena. The advent of instruments such as the scanning tunnelling microscope and the ability of modern computers to give graphical displays of large systems of particles means that a direct connection to nature on the atomic scale has been established.

The challenge of this approach is twofold. The simulationists should aim to bring their computer models closer to experiment whilst the experimentalists should regard the results of the calculations as challenges to prove or disprove. Examples in this chapter have been given in which results established by one methodology have later been verified by another. The depth of origin of sputtered particles described in Section 6.13.2 is one example of a result established by computer simulation that was later confirmed by experiment. The study of bumps caused by individual particle impact on graphite is an example in which simulation has been used not only to predict the formation of the bumps but also to identify the mechanisms responsible for their formation.

It is the view of the authors that the use of MD simulations will continue to grow until they have become an indispensible tool in all surface science laboratories.

References

Allen, M. P. and Tildersley, D. J. (1987). *Computer Simulation of Liquids*, Clarendon Press, Oxford.

Anderson, H. C. (1983). *J. Comput. Phys.* **52** 24.

Bacon, D. J., and Diaz de la Rubia, T. (1994). *J. Nucl. Mater.* **216** 275.

Beeler, J. R. (1985). *Radiation Effects Computer Experiments*, North-Holland, Amsterdam.

Behrisch, R. (1981). *Sputtering by Particle Bombardment*, Springer-Verlag, Berlin, 177.

Berendson, H. C., Postma, J. P. M., Van Gunsteren, W. F., Di Nola, A. and Haak, J. R. (1984). *J. Chem. Phys.* **81** 3684.

Caro, A. and Victoria, M. (1989). *Phys. Rev* A **40** 2287.

Chang, C. C. and Winograd, N. (1989). *Phys. Rev.* B **39** 3467.

Channel, P. J. and Scovel, J. C. (1990). *Nonlinearity* **3** 231.

Evans, D. J. and Morriss, G. P. (1983). *Chem. Phys.* **77** 63.

Firsov, O. B. (1959). *Sov. Phys.: J. Exp. Theor. Phys.* **36** 1076.

Forest, E. and Ruth, R. D. (1990). *Physica D* **77** 63.

Garrison, B. J., Winograd, N., Deavon, D. M., Reimann, C. T., Lo, D. Y., Tombrello, T. A., Harrison, D. E. and Shapiro, M. H. (1988). *Phys. Rev.* B **37** 7197.

Halton, J. H. (1960). *Numerische Mathematik* **2** 84.

Hamers, R. J., Köhler, U. K. and Demuth, J. E. (1990). *J. Vac. Sci. Technol.* A **8** 195.

Harrison, D. E. (1988). *CRC Crit. Rev. Solid St. Mater. Sci.* **14** S1.

Heermann, D. W. (1990). *Computer Simulation Methods in Theoretical Physics*. Springer-Verlag, Berlin.

Jackson, D. P. (1974). *Rad. Eff. Defects Solids* **18** 185.

Lambert, J. P. (1991). *Numerical Methods for Ordinary Differential Systems*, Wiley, Chichester.

Press, W. H., Teukolsky, S. A., Vetterling, W. T. and Flannery, B. P. (1992). *Numerical Recipes*, Cambridge University Press, Cambridge.

Radi, G. (1970). *Acta. Cryst.* A **26** 41.

Ryckaert, J. P., Ciccotti, G. and Berendsen, H. J. C. (1977). *J. Comput. Phys.* **23** 327.

Sanz-Serna, J. M. (1992). *Acta Numerica* **1** 243.

Smith, R., O'Connor, D. J. and von-Nagy Felsobuki, E. I. (1993). *Vacuum* **44** 311.

Smith, R. and Webb, R. P. (1993). *Proc. Roy. Soc.* A **441** 495.

Smith, R. and Harrison, D. E. (1989). *Computers in Physics* September/October, 68.

Smith, R., Harrison, D. E. and Garrison, B. J. (1989). *Phys. Rev.* B **40** 93.

Smith, R., Harrison, D. E. and Garrison, B. J. (1990). *Nucl. Instrum. Meth.* B **46** 1.

Snowdon, K. J., O'Connor, D. J. and Macdonald, R. J. (1988). *Phys. Rev. Lett.* **61** 1760.

Srivastava, D., Garrison, B. J. and Brenner, D. W. (1989). *Phys. Rev. Lett.* **63** 302.

Verlet, L. (1967). *Phys. Rev.* **159** 98.

Webb, R. P., Harrison, D. E. and Barfoot, K. M. (1985). *Nucl. Instrum. Meth.* B **7/8** 143.

Webb, R. P., Smith, R., Dawkaski, E., Garrison, B. J. and Winograd, N. (1993). *Int. Video J. Eng. Res.* **3** 63.

Wilson, I., Knipping, V. and Tsong, I. S. T. (1988). *Phys. Rev.* B **38** 8444.

Zalm, P. C. (1983). *J. Appl. Phys.* **54** 2660.

9

Surface topography

9.1 Introduction

Surface topographic changes occur as a result of all particle–solid inter-actions. For large doses $> 10^{17}$ particles/cm^2 of energetic (tens of kilo-electronvolts) particles, these changes are often visible with the naked eye. Many mechanisms give rise to these topographic features, for example bombardment-induced defects in a solid or differential sputtering yields across a surface, due to grain boundaries or impurity inclusions. Electron micrographs of such surfaces reveal a wide variety of features such as etch pits, ridges, facets, ripples, cones and pyramids. The article by Carter, Navinšek and Whitton (1983) illustrates some of the features that have been observed, but many examples can be found also in most recent issues of the *Journal of Vacuum Science and Technology*. Features that develop can be an unwanted artefact of the ion bombardment technique, for example, they can be responsible for considerable uncertainty in the depth resolution of surface analytical techniques such as dynamic SIMS (secondary ion mass spectroscopy).

On the other hand, surface engineering seeks to etch well-defined patterns on surfaces. Ion beam lithography uses focused beams to etch patterns directly into a substrate but for most technological applications broad beams with masks are used. The modelling of the development of surface shape as a result of particle bombardment is therefore important from the point of view of understanding the basic physical processes involved. In addition it also has important engineering applications.

This chapter will be concerned with developing models that can explain how surface topography changes on the macroscopic scale, both as a result of sputtering (erosion) and deposition or redeposition (growth). The atomistic approach of some of the previous chapters is abandoned in favour of

284

statistical and continuum models that make more general assumptions about the way in which events caused by the impact of a large number of individual particles can be modelled.

The statistical models concentrate on assuming some simple rules each time a particle hits a surface, i.e. certain events occur with a pre-defined probability. These rules are then included in a Monte Carlo simulation. The simplest model will be to consider the substrate as consisting of a regular cellular structure. Deposition occurs as a result of adding cells to a surface and erosion by their removal. If the rules for adding or removing cells from the surface are to be specified in terms of the local environment of each cell then these models are known as 'cellular automata' and the algorithms are easily parallelised for large-scale systems. In a true cellular automaton all the cells are updated simultaneously. If the cells are updated sequentially with a pre-defined statistical probability then this is a Monte Carlo simulation. In this chapter, examples of both systems are given.

The continuum models are based on partial differential equations, which are solved either analytically or numerically to determine the shape of the evolving surfaces. The simplest continuum model is to regard the surface as an advancing wavefront whose speed at any point is known as a function of the bombarding particle and substrate species, the local surface properties and time. For example, in the case of a uniform beam sputtering a homogeneous substrate, the shape of the eroding surface can be calculated if the sputtering yield Y is known as a function of the angle of incidence, θ, assuming that the Y–θ-dependence is the dominant mechanism. Other effects such as surface diffusion, particle reflection and redeposition can also be included in the models if required. Generally, however, it is the sputtering-yield-dependences on target species, crystallographic orientation and incidence angle which are the most important and dominant mechanisms. Much of the understanding of how these features develop has been due to the work of Carter, see for example Carter, Navinšek and Whitton (1983). The theory was developed in the early 1980s, see for example Smith and Walls (1980) and Smith, Carter and Nobes (1986), although this theory seems to have been 'rediscovered' and republished by various authors over the years.

9.2 Cellular models of deposition and erosion

Cellular statistical models can be defined with various degrees of sophistication. In this section we examine some of these models and their predictions. One of the simplest models is to assume that the substrate consists of particles located in cubic (three-dimensional) or square (two-dimensional) cells. Figure 9.1 gives a schematic representation of a two-dimensional square lattice. A surface cell is defined as one that has a missing neighbour and only surface cells are assumed to be eroded. Thus the shaded cell in Figure 9.1(a) is defined as a surface element if any of the adjacent cells number 1, 2, 3 or 4 is missing and that in Figure 9.1(b) if any one of 1–6 is missing. The removal probability of this shaded cell increases with the number of missing neighbours since, if we loosely regard it as an atom, then the fewer neighbours it has, the less strongly bound it will be and we have seen from the transport theory chapter that the ejection yield is inversely proportional to the surface binding energy. It might therefore reasonably be postulated for the model defined in Figure 9.1(a) that a cell with one missing neighbour is removed with probability $p = \frac{1}{4}$, with two missing neighbours $p = \frac{1}{2}$, with three, $p = \frac{3}{4}$ and if it has four missing neighbours then it is not bound and is removed with probability $p = 1$. This is a special case of the power law relation with $n = 1$:

$$p = \left(\frac{b}{4}\right)^n, \quad n = 0, 1, 2 \ldots . \tag{9.1}$$

Here b is the number of missing neighbours and n is the assumed power-law-dependence. This model was initially proposed by Blonder (1986). Thus a simple algorithm to model erosion could be the following.

(1) Set up a two-dimensional array of square cells that defines the substrate and the surface with periodic boundary conditions.
(2) Randomly choose a surface cell.
(3) Remove that cell with a probability given by equation (9.1).
(4) Add new exposed particles to the surface.
(5) Repeat the steps from (2) onwards.

Numerical experiments show a number of features. An initially flat surface remains flat but develops stochastic noise (surface roughness), which stabilises after removal to a depth of about 30 cells. As n increases the surface becomes locally smoother.

In order to demonstrate that the method may be applied to practical problems, two examples are chosen. The first, shown in Figure 9.2(a)–(c), is the erosion of a rectangular structure, which might represent a wire in microelectronic device manufacture. In Figure 9.2(d)–(f), the situation is reversed with deposition (adding cells, rather than removing them) in a rectangular groove. This might represent the planarisation process in microelectronic device manufacture, whereby multilayer devices are made

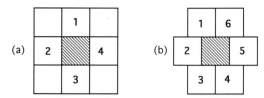

Fig. 9.1 Possible cellular structures for use in two-dimensional simulations. (a) A regular square lattice, in which the shaded cell has four neighbours. (b) An offset square lattice equivalent to a hexagonal lattice in which the shaded cell has six neighbours.

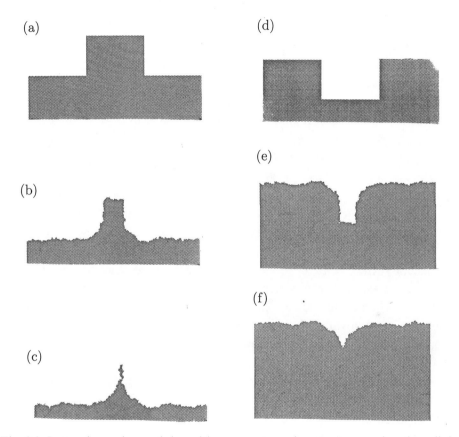

Fig. 9.2 Isotropic erosion and deposition on rectangular structures using the cellular model with $n = 2$: (a)–(c) erosion of a rectangular profile and (d)–(f) deposition in a groove.

by depositing dielectric material in circuits to flatten them before the next structure is laid down.

Figure 9.2 was obtained using the four-neighbour model of Figure 9.1(a) and a value of $n = 2$. The probability of deposition at a site was chosen according to

$$p = 1 - \frac{b}{4}. \tag{9.2}$$

In Figure 9.2, the cell size was chosen as the pixel size on the computer screen in order to obtain good visual images. These figures correspond to isotropic erosion but in many practical applications the erosion is anything but isotropic. One of the most common reasons for the development of surface topography on initially flat surfaces is the inhomogenity of the substrate. Figure 9.3 shows how topography can be generated on an initially flat surface if a stratum of material that might represent a different crystal grain or a material inclusion has a different erosion rate from its surroundings. The simulations in Figure 9.3 were again carried out assuming that $n = 2$ and with the relative erosion rates for the two media given by $c_1/c_2 = 0.5$, where c_i represents the erosion speed of species i. As the erosion continues a shape develops that consists of a curved front edge, a flat top of 'soft' material and a trailing slope at a well-defined angle. The simulation is sufficiently general that it could also be used as a model of a geomorphological process (Smith, 1992).

These cellular models can be easily extended to model a wide range of phenomena. The dependence of erosion rate on surface orientation can be modelled by weighting the removal probability according to the local incidence angle on the surface. This may be estimated by using the heights of adjacent surface cells to obtain an average value for the surface gradient. The dependence on surface curvature can be considered by calculating the number of adjacent substrate cells within a circle of pre-specified radius and weighting the removal probability in proportion to the number of these cells. (This can sometimes lead to instabilities with features growing from initially flat surfaces.) Also, crystal structure can be more accurately simulated by arranging the cells spatially in the correct crystal structure and weighting the removal probabilities according to the binding energy calculated from realistic potentials.

Clearly, modern computer graphics permits the investigation of a large number of phenomena of this kind. Figure 9.4 illustrates a simulation carried out by randomly removing surface cells that have a missing neighbour in direction 1 of Figure 9.1. Such a sequential layer model of sputtering was proposed by Benninghoven (1971). For a surface string of N atoms, the probability of removal of a single cell is always $1/N$, since the number of cells with a missing neighbour in direction 1 is always $1/N$. The probability, after n events, that r removals have taken place from the same position in

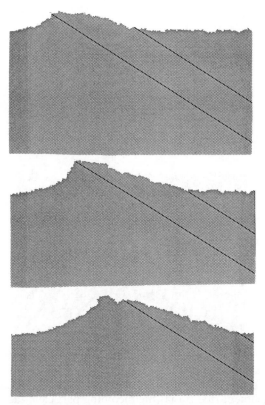

Fig. 9.3 Various stages in the erosion of an initially flat surface composed of two materials or crystal grains with different erosion rates. The thin stratum between the unbroken lines has an erosion rate half that of the surrounding media.

Fig. 9.4 Topography generated from an initially flat surface by the removal of cells with a missing neighbour in direction 1 of Figure 9.1.

the string is

$$P(r) = {}^nC_r \left(\frac{1}{N}\right)^r \left(\frac{N-1}{N}\right)^{n-r}, \tag{9.3}$$

which is the binomial distribution. Letting $n = Nm$ and letting $N \to \infty$ gives

$$P(r, m) = e^{-m} m r / r!, \tag{9.4}$$

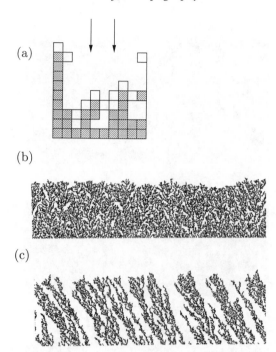

Fig. 9.5 (a) Ballistic deposition on a square lattice at normal incidence. Sites occupied by either the original surface or the growing deposit are shaded and sites where the deposit can grow are shown as unshaded squares. Cyclical boundary conditions are assumed. The attachment rule is that the incoming particle attaches itself to the first site that it encounters on its incoming path. (b) Deposition at normal incidence. (c) Deposition at an angle 62° to the normal. The structures grow at an angle of 35° to the normal.

which is the Poisson distribution. This has mean m and standard deviation $m^{1/2}$. Thus, after removing an average of m atoms from a given position in the surface string, the RMS roughness of the surface is $m^{1/2}$.

This led Benninghoven to propose that the loss of resolution in depth profiling in SIMS after a depth d had been removed was proportional to $d^{1/2}$. However, Figure 9.4 demonstrates the inapplicability of the model because it predicts peaks and troughs that are one cell in width.

There have been many models proposed for ballistic aggregation, which result in fractal structures. Figure 9.5 illustrates a possible model for deposition in which particles are assumed to arrive randomly at the surface and attach themselves to the first adjacent surface cell they meet. As columns begin to grow, arriving particles can either attach themselves to the tops or to the sides of the structures. This can then shadow lower portions of the surface from growth and a deposition develops that consists of particles and holes. The effect is more marked at non-normal incidence. Figures

9.5(b) and (c) compare these two cases. At non-normal incidence a columnar growth takes place at an angle that is different from the angles at which the particles arrive at the surface. There appears to be a correlation between the predictions of these computer models and vapour deposited aluminium films (Meakin *et al.*, 1986).

The examples given above are all Monte Carlo models. An example of a true cellular automaton is provided by the simulations shown in Figure 9.6 for modelling spiral growth at a screw dislocation. The explanation for the spiral growths which can occur at screw dislocations was first given by Frank (1959). Recent atomic force microscopy pictures have confirmed that spirals grown from MOVPE on GaAs consist of steps of single atomic spacing, demonstrating conclusively that spiral growth is an atomic scale phenomenon. A simple cellular automaton algorithm for a square lattice to describe this process is given below. The algorithm is based on the idea that it is difficult to nucleate a new monolayer on a flat surface of an ideal crystal. At each step a new cell is added only at (a) those positions which have an adjacent filled vertical face or (b) at the edge of the dislocation on the next surface, provided that all adjacent cells at the lower level are filled. The second of these mechanisms gives rise to growth on the next layer. The first six steps of this process are shown in Figure 9.6.

9.3 Continuum models of deposition and erosion

9.3.1 Isotropic erosion and deposition

The simplest model of isotropic erosion or deposition is given by Huyghens' wavefront construction. For an isotropic process, all points on the wavefront advance (or are retarded) by a fixed distance d, in the normal direction, in a given time t. For a two-dimensional profile, the new surface after time t is the envelope of all the circular wavefronts drawn from the original surface with radius d. Figure 9.7 shows this for the case of deposition in a square well. This is identical to the situation modelled in Figure 9.2. Note that the convex corners expand into circular arcs whereas concave corners progress as concave corners. This situation is reversed for the erosion process. For deposition on an initially square-well structure the surface rapidly planarises. A simple calculation shows that, if the variation in the level of the surface is Δd after a thickness d has been deposited on the upper flats of the profile then

$$\frac{\Delta d}{d} = 1 - \left[1 - \left(\frac{w}{2d} \right)^2 \right]^{\frac{1}{2}}, \quad d > \frac{w}{2},$$

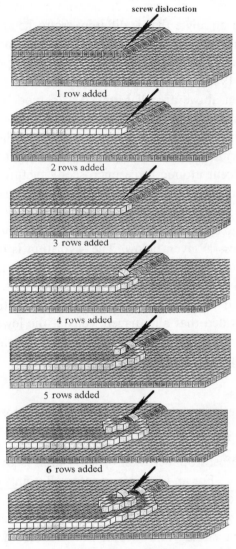

Fig. 9.6 A cellular automaton model of spiral growth at a screw dislocation.

where w is the side length of the square-well structure. For a surface consisting of different media, the Huyghens construction can still be applied. The difference in this case is that the surface erosion rate must be continuous across the boundary between two media. This leads to a law equivalent to Snell's law in optics. For a substrate consisting of material of many different erosion rates, the calculation rapidly becomes complicated and it is easier to study these cases using the Monte Carlo method described earlier.

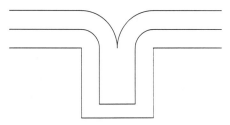

Fig. 9.7 Deposition in a square well using the Huyghens construction.

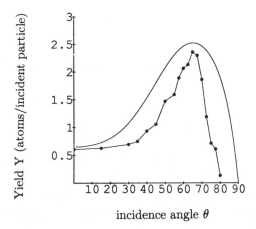

incidence angle θ

Fig. 9.8 The sputtering yield curve as a function of incidence angle for 1 keV Ar bombardment of Si. The smooth curve represents a fit to experimental data and has been taken from Ducommun, Cantagrel and Moulin (1975) ($Y(\theta) = 18.738$ $45\cos\theta - 64.659\ 96\cos^2\theta + 145.199\ 02\cos^3\theta - 206.044\ 93\cos^4\theta + 147.317\ 78\cos^5\theta -$ $39.899\ 93\cos^6\theta$). The points formed by the broken curve are data calculated using MD on the dimer-reconstructed Si$\{100\}(2 \times 1)$ face, see Figure 8.16.

9.3.2 Non-isotropic erosion

Sputter yield variation with incidence angle. The surfaces of amorphous or crystalline materials subjected to inert gas bombardment do not in general erode isotropically. The sputtering yield Y defined here as the number of particles ejected per incoming particle is a function of the angle of incidence θ which the incoming beam makes with the surface normal. The beam flux, J, may also be spatially dependent. For amorphous or polycrystalline materials, Y increases with θ up to a maximum somewhere between $\theta = 40°$ and $\theta = 70°$ and then drops to zero at around 85° incidence, for which reflection without sputtering occurs. Figure 9.8 compares a yield curve numerically fitted to experimental data for 1 keV Ar bombardment of amorphous Si with one calculated using MD on the Si$\{100\}(2 \times 1)$ face, see Chapter 8. Although this yield curve is specific to the Ar–Si interaction, it is

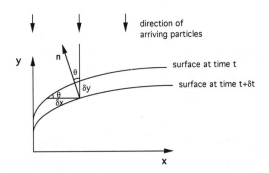

Fig. 9.9 The geometry of the evolution of a two-dimensional contour under erosion by particle bombardment.

typical of many other particle–target combinations and will be used as the basis of the simulations carried out in this section.

The equation which describes the erosion of the surface in two dimensions when the flux J depends on the spatial x co-ordinate is

$$\frac{\partial y}{\partial t} = -\frac{J(x)}{N_A} Y(\theta). \tag{9.5}$$

Here y is the direction opposite to that of the incoming beam and x is the Cartesian co-ordinate perpendicular to y, N_A is the atomic density and t is time. Figure 9.9 illustrates the two-dimensional geometry.

Differentiating (9.5) with respect to x gives

$$\frac{\partial^2 y}{\partial x \, \partial t} = -\frac{J'(x)}{N_A} Y(\theta) - \frac{J(x)}{N_A} Y'(\theta)\frac{\partial \theta}{\partial x} \tag{9.6}$$

and since $\partial y / \partial x = \tan \theta$

$$\frac{\partial \theta}{\partial t} = -\frac{\cos^2 \theta}{N_A} \left(J'(x) Y(\theta) + J(x) Y'(\theta)\frac{\partial \theta}{\partial x} \right). \tag{9.7}$$

If the erosion rate in the x direction is now examined we see that

$$\frac{\partial x}{\partial t} = -\cot \theta \frac{J(x)}{N_A} Y(\theta). \tag{9.8}$$

Similarly, differentiating (9.8) with respect to y gives

$$\frac{\partial \theta}{\partial t} = \frac{J(x)}{N_A} \left[-Y(\theta) + \sin \theta \cos \theta \; Y'(\theta) \right] \frac{\partial \theta}{\partial y}. \tag{9.9}$$

Equations (9.7)–(9.9) are first-order non-linear PDEs of a form originally studied by Lagrange. They can be solved by the method of characteristics. The characteristic lines are given by the solution to the ordinary differential

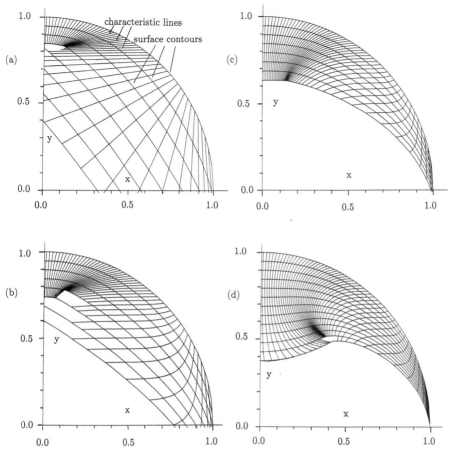

Fig. 9.10 Successive erosion contours of the circle $x^2 + y^2 = 1$ showing also the characteristic lines. The current density is assumed to take the form $J = J_0 \exp(-x^2/\rho^2)$.
(a) uniform beam $\rho = \infty$, (b) $\rho = 1$, (c) $\rho = 1/\sqrt{2}$ and (d) $\rho = 1/\sqrt{3}$.

equations

$$\frac{dx}{dt} = \frac{J}{N_A} \cos\theta \, Y_\theta,$$

$$\frac{dy}{dt} = \frac{J}{N_A} (\cos\theta \sin\theta \, Y_\theta - Y), \qquad (9.10)$$

$$\frac{d\theta}{dt} = -\frac{J_x}{N_A} \cos\theta \, Y,$$

where subscripts denote differentiation. Along the characteristic lines JY/N_A
is a constant. If the flux J is spatially invariant ($J_x = 0$), then the characteristic lines are lines of constant θ and a contour defined by a set of m points
(x_i, y_i, θ_i) $i = 1, 2 \ldots m$ at time $t = 0$ transforms into a set (x_i', y_i', θ_i) at time t

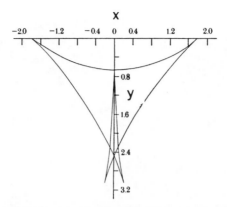

Fig. 9.11 The characteristic locus curve for 1 keV Ar bombardment of Si. The curve is traced out by characteristics emanating from the origin of all surface orientations after a time corresponding to $Jt/N_A = 1$.

according to

$$x_i' = x_i + \frac{J}{N_A} \cos\theta \; Y_\theta t,$$

$$\qquad\qquad (9.11)$$

$$y_i' = y_i + \frac{J}{N_A} (\cos\theta \sin\theta \; Y_\theta - Y)t.$$

The surface shapes are easily calculated once Y and Y_θ are known. The calculated surfaces develop butterfly-like cusps that fold back on themselves. Such cusps are unphysical and care must be taken to remove points that occur in the surface fold. Some results using a Gaussian current density and the Y–θ dependence defined in the caption to Figure 9.8 are shown in Figure 9.10. It can be seen that edges develop from initially smooth surfaces where characteristics intersect and facets form where the characteristics expand.

If the initial surface itself contains an edge (a discontinuity in θ), then this may continue to progress as an edge or a characteristic expansion fan can radiate from the edge. This latter case arises when the characteristics diverge and the surface shape is found from the envelope of the locus of the characteristics of all included angles (this is the same as the edges developing into circular arcs for isotropic erosion, see Figure 9.2). The locus curve for the Y–θ dependence given in Figure 9.8 is given in Figure 9.11. This would be a circle in the case of isotropic erosion. The eroded surface can be drawn from the envelope of the locus curves in a similar way to the Huyghens construction and this forms the basis of some numerical algorithms to solve this problem.

To illustrate the effect of characteristic intersections and expansions, consider the erosion of two planes at right angles bombarded along the negative

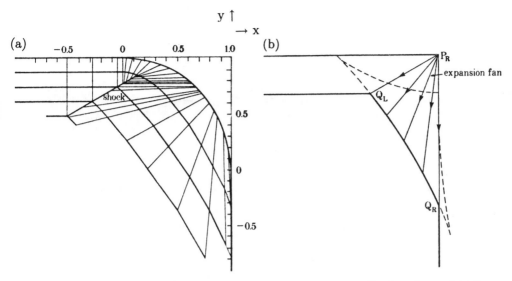

Fig. 9.12 The erosion of a surface section joining two perpendicular planes. (a) The 90° corner, rounded by a circle. The characteristics from the circular section of the surface expand and intersect with those from the normal incident plane and a shock (a discontinuity in θ) develops. A surface gradient discontinuity also forms where the circle intersects the grazing incidence plane $P_R Q_R$. (b) The expansion wave at a 90° corner. The same ultimate shape is produced both in (a) and in (b).

y direction. If the planes are joined by a circular section, then a shock forms due to characteristic intersections. Eventually a smooth section of surface develops, joined by angular discontinuities to the two perpendicular planes. If, on the other hand, the planes meet with an angular discontinuity, then an expansion fan emanates from the corner, as shown in Figure 9.12. Ultimately the end result is the same for both initial conditions. The corner fans out into a smooth portion of surface as before. The angles of orientation of this surface lie between those corresponding to the maximum sputtering yield at the point Q_R and that whose erosion rate in the y direction is the same as for the flat plane, at the point Q_L.

The theory readily extends to three dimensions. In this case, Y can depend on two angles θ and ψ defined in Figure 9.13, which also defines the co-ordinate system to be used. If the surface is given at time t by $S(\mathbf{r}, t) = 0$ then expanding in a Taylor series and taking the limit as $\delta t \to 0$ gives

$$\frac{\partial S}{\partial t} + (\mathbf{c} \cdot \nabla)S = 0, \tag{9.12}$$

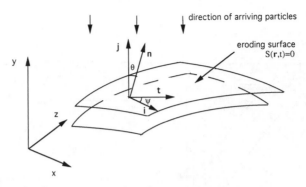

Fig. 9.13 The geometry of a three-dimensional eroding surface.

where c is the normal velocity of the surface. In fact

$$c = \frac{g}{\cos\theta}, \quad g = \frac{J}{N_A}Y.$$

The characteristics for a spatially uniform beam are then given by the equations

$$v_x = \frac{dx}{dt} = \frac{J}{N_A}\left(\cos^2\theta\sin\psi\ Y_\theta + \cos\psi\cot\theta\ Y_\psi\right),$$

$$v_y = \frac{dy}{dt} = \frac{J}{N_A}(\cos\theta\sin\theta\ Y_\theta - Y), \tag{9.12}$$

$$v_z = \frac{dz}{dt} = \frac{J}{N_A}\left(\cos^2\theta\cos\psi\ Y_\theta - \sin\psi\cot\theta\ Y_\psi\right),$$

along which θ and ψ are constants. A derivation has been given by Smith, Carter and Nobes (1986) and for spatially and time-dependent beams by Tagg, Smith and Walls (1986).

Tracing the new surface after time t can be done as in two dimensions. A set of m points (r_i, θ_i, ψ_i) $i = 1, \ldots n$ at time $t = 0$ transform into a set (r'_i, θ_i, ψ_i) at time t according to

$$x'_i = x_i + v_x t; \quad y'_i = y_i + v_y t; \quad z'_i = z_i + v_z t. \tag{9.13}$$

An example that uses this algorithm (removing the folds as before) is shown in Figure 9.14 with $Y_\psi = 0$. This demonstrates how a ridge may form by the erosion of an initially smooth elliptical hummock.

Although the characteristic method gives much physical insight into the erosion process, it does not necessarily convert into a good algorithm for carrying out the calculations. An excellent algorithm (DINESE) has been written by Katardjiev *et al.* (1994) for three-dimensional erosion applications.

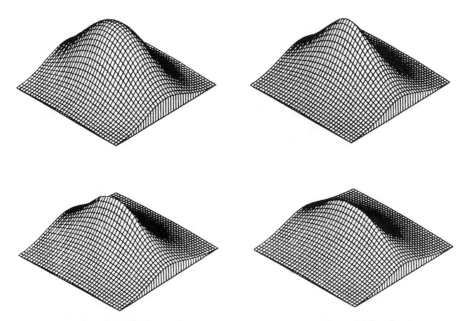

Fig. 9.14 Stages in the erosion of the surface $z = \exp(-x^2 - 0.4y^2)$ after successive equal doses of ions. A sharp ridge has formed by the last stage.

The effect of surface curvature. The dependence of the surface erosion or growth rate on surface curvature has been used to explain the formation of ripple topography on ion-bombarded surfaces. The dependence of the sputtering yield on incidence angle alone cannot do this since such a dependence implies that a homogeneous medium subjected to erosion by a uniform beam would have an ultimately flat surface. Ripple topography can be predicted if surface diffusion is taken into account. It can also form due to the local surface curvature, enhancing or reducing the ejection yield. The way in which this might occur is demonstrated in Figure 9.15. In Figure 9.15 the elliptical lines denote contours of equal energy deposition. The energy deposited at 0 by the ions striking the surface at 0 is the same as that deposited at 0' by the ions striking the surface there. However, the average energy deposited at 0 by an ion that hits the surface at A is greater than that deposited at 0'.

If the depth at which the maximum energy deposition takes place is a and if $aK \ll 1$, where K is the surface curvature, then Bradley and Harper (1988) have shown that the surface recession in two dimensions takes the form

$$\frac{\partial y}{\partial t} = f(\theta) + g(\theta)K, \qquad (9.13)$$

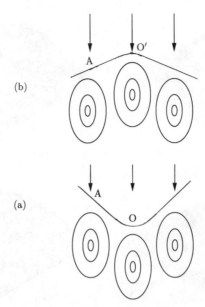

Fig. 9.15 A normally incident ion beam striking (a) a trough and (b) a crest. The arrows indicate the beam direction. The elliptical contours indicate contours of equal energy deposition.

where the functions f and g depend on the form of the energy deposition profile and g is negative near $\theta = 0$. Periodic small disturbances in a surface grow rapidly and surface self-diffusion has to be incorporated in the model in order that realistic wavelengths of the ripple structures are predicted. Assuming that surface diffusion is the only process operating, denote $\mu(K)$ to be the increase in chemical potential per atom to a point of curvature K on the surface. Following Herring (1951)

$$\mu(K) = K\gamma\Omega, \tag{9.14}$$

where γ is the surface free energy per unit area and Ω is the atomic volume. Gradients of chemical potential along the surface will produce a drift of surface atoms with an average velocity given by the Nernst–Einstein relation

$$v_{\mathrm{s}} = -\frac{D_{\mathrm{s}}}{kT}\frac{\partial \mu}{\partial s} = \frac{D_{\mathrm{s}}\gamma\Omega}{kT}\frac{\partial K}{\partial s}, \tag{9.15}$$

where D_{s} is the coefficient of surface diffusion, s is the arc length, T is temperature and k is Boltzmann's constant. The surface current J_{s} is the product of v_{s} and the number of atoms per unit area v:

$$J_{\mathrm{s}} = -\frac{D_{\mathrm{s}}\gamma\Omega v}{kT}\frac{\partial K}{\partial s}. \tag{9.16}$$

The increase in the number of atoms per unit area per unit time is $-\partial J_s/\partial s$ and, by multiplying this expression by Ω, we obtain the expression for the rate of advance along the normal $D_e \, \partial^2 K/\partial s^2$, where $D_e = D_s \gamma \Omega^2 v/(kT)$.

Thus, in the y direction, using $K = y_{xx}/(1 + y_x^2)^{3/2}$, we have

$$\frac{\partial y}{\partial t} = -D_e \frac{\partial}{\partial x}\left[(1 + y_x^2)^{-\frac{1}{2}} \frac{\partial}{\partial x}\left(\frac{y_{xx}}{(1 + y_x^2)^{3/2}}\right)\right]. \tag{9.17}$$

The term on the right-hand side of equation (9.17) can then be incorporated into (9.13) to produce the equation

$$\begin{aligned}\frac{\partial y}{\partial t} = &f(\theta) + g(\theta)\frac{y_{xx}}{(1 + y_x^2)^{3/2}} \\ &- D_e\left\{(1 + y_x^2)^{-1/2}\left[y_{xx}(1 + y_x^2)^{-3/2}\right]_x\right\}_x.\end{aligned} \tag{9.18}$$

If f and g are now assumed to be constant then Srolovitz (1988) has shown by making the substitutions $H = [-g/(2D_e)]^{1/2}(h - ft)$, $z = [-g/(2D_e)]^{1/2}x$ and $\tau = [g^2/(4D_e)]t$ that equation (9.18) becomes

$$H_\tau = -2\left[H_z(1 + H_z^2)^{-1/2}\right]_z - \left\{(1 + H_z^2)^{-1/2}\left[H_z(1 + H_z^2)^{-1/2}\right]_{zz}\right\}_z. \tag{9.19}$$

This has a steady state solution given by

$$H^2 - (1 + H_z^2)^{-1/2} = C, \tag{9.20}$$

where C is an integration constant. For $-1 < C < 0$

$$H = (1 + C)^{1/2}\cos\alpha, \tag{9.21}$$

where α depends on z via

$$2E(\alpha, r) - F(\alpha, r) = \sqrt{2}z + \text{constant},$$

where $r = [(1 + C)/2]^{1/2}$ and F and E are incomplete elliptical integrals of the first and second kind. When $C > 0$ the solution of (9.20) is still given by (9.21) until $\sin^2\alpha$ reaches $1/(1 + C)$. The solution then jumps from $H = C$ to $H = -C$ (or vice-versa). These steady states with $C > 0$ are stable with respect to small perturbations. Some solutions are plotted in Figure 9.16.

The partial differential equations which describe the time-dependent progression of the surface when curvature effects are important are not hyperbolic and cannot be solved by the method of characteristics. Because of the inherent instability of the physical process, finite-difference algorithms are themselves often unstable. Some numerical algorithms that can be used have been given by Sethian (1990).

The model described above, in which ripple structures arise as a result of surface curvature effects modified by diffusion, can also be modelled by

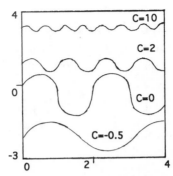

Fig. 9.16 The steady state analytical solutions of the surface evolution equation for various values of the undetermined constant of integration C.

the cellular methods described at the beginning of the chapter. Meakin, Family and Vicsek (1987) proposed a very simple model of cellular growth, which depends on the surface curvature. The curvature is approximated by counting the number of substrate cells within a circle of a pre-set radius (which is adjustable depending on the problem) centred on a surface cell. Diffusion is modelled by allowing particles to hop across the surface. The simulations show that, in the absence of diffusion, unstable columnar growth occurs. Diffusion stabilises this growth.

9.4 Re-deposition

Re-deposition of material ejected by energetic particle bombardment can be important along the edges of steep surface features and can be observed along the edges of microcircuit patterns. Figure 9.17 shows an example whose ridges of re-deposited Si remain when the resist material is dissolved. The re-deposition is due to emitted particles of low energy. In order to model this phenomenon, it is necessary to calculate the secondary flux at a point where the re-deposition is occurring. To do this we must model the angular distribution of the ejected material. The secondary particle emission flux ϕ_1 from the surface at a point depends on the angle θ_E between the direction of emission and the surface normal. This distribution can be modelled by a low-order polynomial in $\cos \theta_E$, but for normal incidence bombardment a simple cosine distribution provides a good approximation, where

$$\phi_1 = \phi_0 \cos \theta_E \tag{9.22}$$

and ϕ_0 is the peak emitted flux density.

Consider the two-dimensional situation shown in Figure 9.18. The surface OX'' is being bombarded by a uniform beam and the point A on the surface

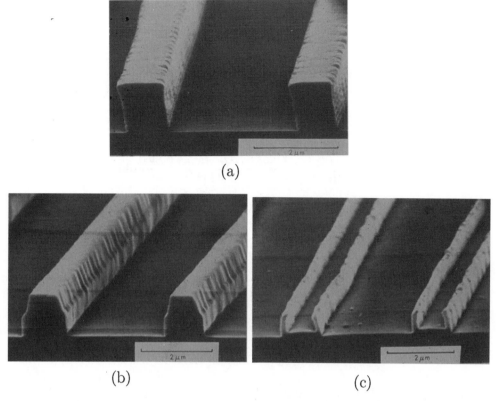

(a)

(b)   (c)

Fig. 9.17 Scanning electron micrographs illustrating the phenomena of facetting and re-deposition during ion etching. The micrographs show a photoresist mask of a Si substrate (a) before ion etching, (b) after etching to a depth of 1300 Å and (c) the re-deposited material after dissolution of the solvent.

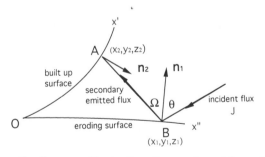

Fig. 9.18 A schematic diagram illustrating the re-deposition of sputtered material from a point B on the eroding surface to a point A.

OX' is receiving a secondary flux of material from the surface elements at B. The flux density of re-deposited material at A from the surface element at B is

$$\delta\phi_2 = \frac{JY\cos\theta\,dz_1\,|ds_1|}{\pi N_A|AB|^4}(AB\cdot n_1)(AB\cdot n_2),\tag{9.23}$$

where ds_1 is the arc length of the surface at B and the direction of z_1 is perpendicular to the symmetry plane. The total flux density at A from a sputtered two-dimensional surface between the lines $s_1 = a$ and $s_1 = b$ is

$$\phi_2 = \frac{J}{\pi N_A}\int_b^a\int_{-\infty}^{\infty}\frac{Y\cos\theta\,(AB\cdot n_1)(AB\cdot n_2)}{|AB|^4}dz_1\,ds_1.\tag{9.24}$$

This flux integral cannot be easily evaluated in general. However, evaluation is possible for the case in which OX'' is a flat plane and then the normal rate of build up v_n at A is given by

$$v_n = \frac{\mu J}{2N_A}\cos\theta\left[1+\left(\frac{dy_2}{dx_2}\right)^2\right]^{1/2}\left(\frac{dy_2}{dx_2}(\cos\beta_2 - \cos\beta_1)\right.$$

$$\left.+\sin\beta_2 - \sin\beta_1\right)\tag{9.25}$$

when μ is the sticking coefficient and

$$\begin{aligned}\beta_1 &= \arctan\left(\frac{x_0 - x_2}{y_1 - y_2}\right),\\\beta_2 &= \arctan\left(\frac{d - x_0 - x_2}{y_1 - y_2}\right).\end{aligned}\tag{9.26}$$

Here d represents the initial groove width and x_0 the value of x corresponding to the lower limit for the s_1 integration in (9.24).

For a surface whose normal velocity is given by equation (9.25) the method of characteristics can again be applied to produce a solution. The characteristics are the lines

$$\begin{aligned}\frac{dx_2}{dt} &= \frac{\mu JY\cos\theta}{2N_A}(\cos\beta_2 - \cos\beta_1),\\\frac{dy_2}{dt} &= \frac{\mu JY\cos\theta}{2N_A}(\sin\beta_2 - \sin\beta_1).\end{aligned}\tag{9.26}$$

Some computed results for different groove widths are presented in Figure 9.19. A fuller discussion has been given by Smith, Makh and Walls (1983).

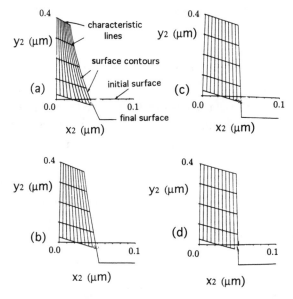

Fig. 9.19 Theoretically calculated redeposition profiles for rectangular grooves of different widths and for different depths of erosion using the cosine distribution and assuming $\mu = 1$. The depth of erosion corresponding to each surface contour is 10 nm and the groove height is 0.4 μm. Groove widths are (a) 0.5 μm, (b) 1.0 μm, (c) 3.0 μm and (d) 6.0 μm.

9.5 Other secondary effects

In the ion bombardment of solids, the primary effect which causes surface topography to develop and change is the differential sputtering yield of the various substrate constituents combined with the dependence of the yield on incidence angle. Curvature effects and re-deposition are generally secondary phenomena in comparison.

Another secondary phenomenon that has been modelled is the effect of ion reflection, which can be a problem near steep-sided surface structures in microelectronic device fabrication. A full continuum description of all these phenomena is possible but the full equations are very difficult to solve. This means that general Monte Carlo models based on the cellular approach such as those of Rossnagel and Robinson (1983) provide the most convenient approach.

9.6 Conclusion

The continuum models can be used to predict the surface topography which can occur on a macroscopic scale as a result of ion bombardment. For a uniform ion beam for which the surface erosion rate depends only on the surface orientation, any smooth undulations in the surface can develop into edges, but are ultimately smoothed out for large times. If the erosion rate depends on the surface curvature, then it is possible for non-planar structures to develop, which are stable for long timescales. It has been shown in Chapter 8 how single-particle impacts can produce craters or bumps on a surface but so far it has not been possible to run MD simulations for multiple particle impacts on realistic systems in order to observe the development of topographic features. At the present time, therefore, there is a gap in the modelling between the microscopic description inherent in the MD simulations and the macroscopic predictions of the continuum models. It is expected that this gap will narrow during the next few years as computing capabilities improve so that the atomistic models become more realistic and the continuum models are improved to include more physical effects.

References

Benninghoven A. (1971). *Z. Phys.* **230** 403.

Blonder, G. E. (1986). *Phys. Rev.* B **33** 6157.

Bradley J. and Harper, M. E. (1988). *J. Vac. Sci. Technol.* A **6** 2390.

Carter, G., Navinšek, B. and Whitton, J. L. (1983). *Sputtering by Particle Bombardment II* Ed. R. Behrisch, Springer-Verlag, Berlin, ch 6.

Ducommun, J. P., Cantagrel, M. and Moulin, M (1975). *J. Mater. Sci.* **16** 52.

Frank, F. C. (1959). *Growth and Perfection of Crystals*, Wiley, New York.

Herring, C. (1951). *Physics of Powder Metallurgy*, McGraw-Hill, New York.

Katardjiev, I. V., Carter, G., Nobes, M. J., Berg, S. and Blom, H. O. (1994). *J. Vac. Sci. Technol.* A **12** 61.

Meakin, P., Ramanlal, P., Sander, L. M. and Ball, R. C. (1986). *Phys. Rev.* A **34** 5091.

Meakin, P., Family, F. and Vicsek, T. (1987). *J. Colloid Interface Sci.* **177** 394.

Rossnagel, S. M. and Robinson, R. (1983). *J. Vac. Sci. Technol.* A **1** 426.

Sethian, J (1990). *J. Diff. Geom.* **31** 131.

Smith, R., Makh, S. S. and Walls, J. M. (1983). *Phil. Mag.* **47** 455.

Smith, R. and Walls, J. M. (1980). *Phil. Mag.* **42** 235.

Smith, R., Carter, G. and Nobes, M. J. (1986). *Proc. Roy. Soc.* A **407** 405.

Smith, R. (1992). *Earth Surf. Processes Land Forms* **16** 273.

Srolovitz, D. J. (1988). *J. Vac. Sci. Technol.* A **6** 2371.

Tagg, M. A., Smith, R. and Walls, J. M. (1986). *J. Mater. Sci.* **21** 123.

Index

307

1-MONTH